Advanced Practical Inorganic and Metalorganic Chemistry

T0318207

Advanced Practical Inorganic and Metalorganic Chemistry

R.J. Errington

Department of Chemistry,
University of Newcastle upon Tyne,
Newcastle upon Tyne, UK

BLACKIE ACADEMIC & PROFESSIONAL
An Imprint of Chapman & Hall
London · Weinheim · New York · Tokyo · Melbourne · Madras

Published by Blackie Academic & Professional, an imprint of Chapman & Hall, 2–6 Boundary Row, London SE1 8HN, UK

Chapman & Hall, 2–6 Boundary Row, London SE1 8HN, UK

Chapman & Hall GmbH, Pappelallee 3, 69469 Weinheim, Germany

Chapman & Hall USA, 115 Fifth Avenue, New York, NY 10003, USA

Chapman & Hall Japan, ITP-Japan, Kyowa Building, 3F, 2-2-1 Hirakawacho, Chiyoda-ku, Tokyo 102, Japan

DA Book (Aust.) Pty Ltd, 648 Whitehorse Road, Mitcham 3132, Victoria, Australia

Chapman & Hall India, R. Seshadri, 32 Second Main Road, CIT East, Madras 600 035, India

First edition 1997

© 1997 Chapman & Hall

Typeset in 10/12pt Times by Acorn Bookwork, Salisbury, Wiltshire
Printed in Great Britain by St Edmundsbury Press Ltd, Bury St Edmunds, Suffolk

ISBN 0 7514 0225 7

A catalogue record for this book is available from the British Library
Library of Congress Catalog Card Number: 97–71914

∞ Printed on acid-free text paper, manufactured in accordance with ANSI/ NISO Z39.48-1992 (Permanence of Paper)

To Ruth

Contents

CONTENTS

Preface

Several years ago, it became clear to me that students and supervisors alike would benefit from an affordable text which described a range of techniques likely to be encountered during a research project in inorganic chemistry. Discussions with others working in the area reinforced this idea, and when Jon Walmsley of Blackie Publishers approached me with a view to producing such a text, I agreed to take on the task.

That was longer ago than I care to remember, and Jon has seen a few test series come and go in the meantime. The resulting book has been written for postgraduate and advanced level undergraduate chemistry students embarking upon inorganic and metalorganic research, and will also be a useful reference source for materials scientists in university and industry.

I have tried to emphasize the need for a thorough approach to practical inorganic and metalorganic chemistry (i.e. the chemistry of metal and metalloid compounds, including those which incorporate organic fragments), and to show how ideas and techniques from other areas can provide opportunities for new developments. As always in a book of this nature, it has been necessary to balance the breadth of scope against the amount of detail presented, so references for further in-depth reading are provided throughout.

Although this book was a solo effort in terms of preparation, the help I've received from various individuals has been invaluable. My thanks go to all those who read parts of the manuscript and offered much needed encouragement: Mark Winter, Paul Pringle, Nick Norman and, in particular, Todd Marder, who read the whole manuscript and made several useful suggestions. William McFarlane, Andrea Russell and Robin Perutz provided constructive comments on the NMR, electrochemistry and matrix isolation sections, respectively.

I particularly want to mention all those, in chronological order, who have encouraged and enabled me to make a career out of a rather exotic hobby: my parents (for allowing me to build a chemistry laboratory in the back yard), Dennis Arthur (my chemistry teacher at what was then King James I Grammar School, Bishop Auckland), Bernard Shaw (Leeds University), Malcolm Chisholm (Indiana University) and Don Bradley (Queen Mary College).

Thank-you, Apple^{TM} for producing computers I can live with (literally). For those who are interested, the text was prepared using Microsoft

Word™ and diagrams were drawn with Chemintosh™ on a Quadra 660AV, a PowerBook 160 and a PowerMac 7500.

Special thanks, however, are reserved for Ruth, for all her understanding and encouragement during the stressful times.

John Errington
Newcastle upon Tyne

1 General Introduction

A research training in preparative inorganic chemistry should instil self-confidence derived from an ability to handle substances with diverse properties, e.g. reactive gases, volatile and pyrophoric reagents, moisture-sensitive compounds and refractory solids. Undergraduate laboratory classes provide the fundamentals of synthetic chemistry, but time and resource limitations restrict the treatment of more advanced techniques. A final year undergraduate project in a research laboratory may provide first-hand experience of more sophisticated experimental methods but, again, time is often limited.

This book has been written to bridge the gap between teaching and research environments, to help new research students find their feet in the laboratory, and to introduce techniques for handling air-sensitive compounds. Practical chemistry is necessarily a very personal experience and, while the basis for this text is naturally my own research experience and interests, I have tried to combine coverage of the basic techniques with a description of selected, more specialized procedures (including some which have only recently been adopted by inorganic chemists) in an effort to provide beginners and non-specialists with the confidence to tackle the type of preparative work they are likely to encounter in the literature.

Although specific areas of chemistry are not covered in detail, the techniques described are generally applicable to reactions involving air-sensitive compounds. Similarly, the discussion of spectroscopic techniques is limited to the practical aspects, and suitable references are provided where more detailed information may be required.

For the illustrations throughout the book, I have chosen to use examples of apparatus incorporating screw joints. While we have found them to be very convenient and to have advantages over ground glass in many instances, ground-glass joints may, of course, be substituted in the majority of cases.

In chapter 13, rather than describe the preparation of a few selected compounds in detail, I have summarized preparative methods for several classes of compound which are useful as starting materials. Wherever possible, tested methods from *Inorganic Syntheses* are provided, but references to significant compounds from the recent primary literature are also included. This book provides the background to enable competent researchers to repeat all of the syntheses in this chapter.

The aim of this guide is, therefore, to provide newcomers to the field with:

- the information necessary to get started in the laboratory;
- rapid access to the accumulated wealth of practical expertise in synthetic inorganic/metalorganic chemistry which is available in the literature;
- the confidence to develop improvements and try out their own ideas.

Hopefully, experienced bench chemists should also find something of interest within these pages. We should all remain alert to alternative methods and be aware of any new developments in synthetic techniques.

2 Preliminaries

2.1 The laboratory

For a graduate student joining a research group there is usually an initial feeling of strangeness in the unfamiliar laboratory, and even postdoctoral researchers who have worked in the area and who have some experience of the general apparatus and techniques take a while to adapt to new surroundings. This is worth remembering while you search every cupboard and drawer in the lab for that one particular item of glassware or apparatus you need to set up your first reaction. Don't hesitate to ask established postgrad or postdoc co-workers (or even your supervisor!) about the organization of the lab and the department (it is surprising how many people feel inhibited or don't wish to appear helpless, and as a consequence get little done in the first few weeks).

Research groups evolve slightly different protocols for the day-to-day routine, but the essential elements are common to all those engaged in synthetic work and, for the purposes of this discussion, I have divided these into the facilities required and some procedures with which you need to become familiar (Scheme 2.1). In later chapters I will discuss various techniques in some detail, but the purpose of this section is to provide an overview and comment on general aspects of lab management.

2.1.1 Bench space

Most work will be carried out in what you will come to regard as your own bench space or in a fume cupboard. Around the lab there will also be communal bench space for shared equipment, and the amount of space allocated to you will clearly depend on the number of people in the lab and the amount of communal equipment. Nowadays, it is standard practice to carry out preparative metalorganic reactions under an inert atmosphere so you will need the necessary facilities (chapter 3) at your bench in order that oxygen and moisture can be routinely excluded from reactions. Larger, permanent items of glassware are best supported on a lattice framework of metal or fibreglass rods which is fixed firmly to the bench and/or a wall (Figure 2.1). Although construction is fairly straight-forward, several features can make life at the bench easier and these should be borne in mind from the outset. For electrical safety, a metal framework should be connected to an earth point. The rods should be

Facilities
(individual and communal) {
Bench space
Fume cupboards
Apparatus and glassware
Solvent stills

Procedures
(individual and communal) {
Procuring chemicals and equipment
Carrying out reactions
Cleaning and drying glassware
Disposal of residues
Using departmental facilities

Scheme 2.1   Essential elements of laboratory work.

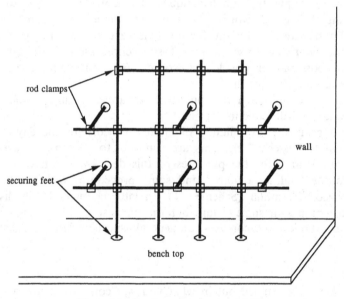

Figure 2.1   Bench-mounted support lattice.

positioned so as not to obstruct access to electrical sockets and switches
or to gas taps and shelves situated behind the framework. Feet securing
the rods to the bench top should be as small as possible and the circular,
three-hole types can be sawn down for attachment by a single screw.
Horizontal rods are included mainly to impart rigidity to the framework;
the weight of any items clamped to these rods can cause the clamps to
rotate downwards. Apparatus should, therefore, be clamped to the
vertical rods which can be arranged conveniently across the working area
and be of sufficient height to support permanent items such as gas purification columns and vacuum manifolds, which need to be positioned so as

not to obstruct your work at the bench. Always try to organize the bench top so that it is easy to keep clean and tidy and so that apparatus can be used with the minimum of movement on your part. If possible, a separate area of desk space should be available where you can write up your lab book and keep spectra and other paperwork.

2.1.2 Fume cupboards

In an ideal world, all reactions would be carried out in efficient fume cupboards, but financial and space restrictions mean that in academic labs these are normally shared, so reactions involving less hazardous substances are often carried out at the bench. This does not normally pose any problems when carrying out reactions in a controlled environment, since the aim is to protect reagents from the atmosphere, but see section 2.4 which deals with safety considerations. Fume cupboard performance should be monitored regularly using a rotating-vane anemometer and the details recorded in a notebook. Cotton threads or a piece of tissue paper taped to the lower edge of the sash provide a permanent visual indication of air-flow. Note that a reduction in efficiency or even fan failure can result if used pieces of paper towel, aluminium foil or cotton wool are drawn into the extract system so, as a matter of routine, these should always be removed from fume cupboards.

2.1.3 Apparatus and glassware

Details of glassware and equipment for routine use are given in chapter 3. You will probably acquire your own personal collection, but much will be communal and a certain amount of discipline is required from everyone in the lab to keep shared equipment clean and in good condition. A high-vacuum line will probably be shared, as will more frequently used equipment such as dry boxes, ovens, fridges, freezers and balances. Everyday electrical items such as stirrers, heater/stirrers, heating mantles and hot air blowers (of the 'hair dryer' or 'paint-stripper' type) need to be maintained in good working order and checked regularly for frayed, worn or melted cables.

2.1.4 Solvent stills

Dry, oxygen-free solvents are needed for most reactions, and communal solvent stills are usually set up in the lab for the commonly used solvents. To function properly, stills need frequent attention and maintenance and are a potential source of friction within the group. It is everyone's responsibility to ensure the correct use and proper maintenance of the stills, and one good way to achieve this is to organize a rota for cleaning and regeneration.

2.1.5   Procuring chemicals and equipment

From time to time, you will find that certain chemicals or items of glassware are not available in the lab and you will need to know how to buy new supplies. This is normally done centrally within a department and, after identifying a suitable supplier, you will probably have to fill out an order form and hand it to someone in the stores. This should be done in consultation with your supervisor, firstly to ensure that funds are available, and secondly to check that disposal procedures for all residues and unused chemicals can be followed, since this can often cost more than the chemicals themselves.

2.1.6   Carrying out reactions

The details of each experiment should always be planned carefully and it often helps to sketch apparatus before assembling it. Try to carry out the experiment mentally first, so that you can envisage each step and antici-pate any potential problems with material transfer or manipulation. Synthetic chemistry is an art, and synthetic chemists are usually judged by their ability to choose appropriate reagents and reaction conditions for the preparation of new compounds. Talk to most people who have been in the game for some time and you will find that they are all in it for the same reason – they enjoy making new compounds, and it is often this enthusiasm which distinguishes the outstanding research worker from the rest. As you gain confidence and become more familiar with the techni-ques, setting up the glassware for reactions will become second nature to you, but you should never allow yourself to become complacent.

Carrying out a reaction is often the easiest part of synthetic work. Much more effort is usually involved in the work-up, i.e. the separation and purification of the various products, and in the subsequent characteri-zation of compounds obtained, and you should be prepared for a good measure of frustration as you tackle these fundamental challenges. Your reward for all your efforts will in the main part be the sense of elation which accompanies success.

2.1.7   Cleaning and drying glassware

This is a very important part of lab work and it is essential that you are able to do it properly since there isn't much point in using dirty glassware for your reactions. A poor performance in this area when it comes to shared apparatus is likely to upset other group members, so it is best to get it right from the beginning! After rinsing with an organic solvent such as acetone or industrial ethanol (methylated spirits), flasks can be washed thoroughly with hot water and detergent using a test-tube brush to remove solids from the glass surface. Stubborn residues can usually be removed with care using concentrated nitric acid (make sure that no alcohol remains in the flask, otherwise the vigorous reaction may spray acid into your face) or alkaline peroxide solution (for early transition-metal oxide residues). Chromic acid baths have traditionally been used to remove organic residues, but these are best avoided since chromium salts are now listed as carcinogens and they can also destroy essential bacteria in water treatment plants. In some cases, an ethanolic KOH bath can be used to soak glassware overnight. This will help to remove residual silicone greases, but items should not be left in the bath for too long, as the glass will gradually dissolve and become weakened (especially hazardous for glassware that is to be evacuated). This applies especially to sintered glass filters and to ground-glass joints and taps.

Once the glassware is clean it can be dried in an oven at 120–150 °C. It is advantageous if you are to be working with moisture-sensitive compounds to use ethanol (industrial meths) for the final rinse since it forms an azeotrope with water which helps to remove residual moisture. Dry glassware should then be assembled quickly while hot (wear protective gloves) and allowed to cool under vacuum to minimize exposure to atmospheric moisture. If the glassware is not to be used for a while, it should be stored in a cupboard or drawer and protected from impact with round-bottom flasks, which tend to roll around and cause 'star' cracks when drawers are opened and closed too vigorously. It is, therefore, a good idea to line drawers with cotton wool or a thin layer of foam packing which can be fixed to the bottom of the drawer with double-sided adhesive tape.

2.1.8 Disposal of residues

In the UK, anyone using chemicals, whether in industry or a university, has a legal responsibility to ensure that all waste is disposed of safely. When planning experiments, the final fate of all solvents, reagents and products must be borne in mind from the outset and a safe means of their disposal must be available. Only non-toxic aqueous wastes should be poured down lab sinks. Acids and bases should be highly diluted, and it is best to avoid disposing of any organic solvent down the sink, although small amounts of alcohol from the washing of glassware pose no hazard if diluted with water. Organic solvent residues should be collected and given to a registered waste disposal company whose advice will determine which residues can be combined. Acetone can be recycled if it has been used only for rinsing cleaned glassware and can, therefore, be collected separately. However, solvents used in reactions will be contaminated with a wide variety of other materials and cannot be recycled and will probably be incinerated. Chlorinated solvents require high temperature incineration and need to be collected separately along with any mixed solvents in which they are present. Flammable non-chlorinated solvents are often used as fuel for incinerators and can be collected together. Any particularly toxic solvents should be collected separately and labelled appropriately prior to disposal. At all times, it is essential to take precautions to avoid the generation of potentially reactive mixtures and, if you are in any doubt, store the residues separately and label them properly.

2.1.9   Using departmental facilities

Throughout your time in the lab, you will need to use the support services provided by the department. This means that you will have to develop good working relationships with a variety of people including cleaners, technicians and secretaries in addition to other researchers. Always remember that you are dealing with skilled people who take pride in their job, have been working in the department for some time, and who probably will still be there after you move on. The services of a good glassblower or an experienced technician from the mechanical or electronics workshops are invaluable, but you should bear in mind that you are not the only one with urgent requirements. The same applies to other technicians, including those responsible for analytical and spectroscopic services. The bottom line is that if you treat all these people with due respect, you will receive a much better service in return and may learn a few things in the process. Take time to find out the correct procedure for submitting jobs or samples and always be prepared to discuss individual requirements with support staff; they will often be able to suggest useful modifications to apparatus design or perhaps propose alternative spectroscopic experiments.

2.2   The literature

The first step of any research project should always be a literature search. You need to be aware of the background to the project, identify others working on the same or related topics and establish the current state of knowledge in the area. Regular visits to the library should become part of your routine so you can stay abreast of any developments by scanning current journals and reviews. It will take a little time to recognize which journals are likely to contain material most relevant to your work, but once this has been established these will form the basis for your regular visits to the current periodicals section of the library. The easiest way to find out which journals to look at and where to find them is to ask your colleagues; they will have already been through this process and accumulated their own expert knowledge. In addition to the 'main-line' journals, there are those of a more general, applied or interdisciplinary nature and you will probably also want to maintain an awareness of major developments in other areas of chemistry. The following sections on literature use and information retrieval are intended to help you make full use of the excellent facilities which are now readily accessible.

2.2.1 Finding your way through the chemical literature

Your first confrontation with the assembled chemical literature, probably as an undergraduate, may well have been a daunting experience. The rows of bound volumes are an indication of the vast amount of stored information, and the output of the world's chemical community is increasing all the time. Approximately 12 million papers, patents, reports, etc. have been published to date, and this is expected to double over the next 15 years. An understanding of how the published material is organized will certainly minimize frustration as you ease your way through the labyrinth, and I recommend that you look at one of the detailed guides to information sources [1–3]. One of the most recent [1] is very readable and includes coverage of developments that have occurred in the last few years.

Results from research are usually published first in the primary literature, i.e. as articles in journals, in theses or in patents. The list in Table 2.1 is by no means comprehensive, but gives some idea of the journals which are likely to be of interest. A means of accessing the vast amount of primary material is provided by several abstracting and indexing publications which constitute the secondary literature. These appear soon after the original article and are to be found in the reference section of the library. Perhaps the most important (and expensive) publication of this type is *Chemical Abstracts* (CA), the only one which aims to give comprehensive coverage of all chemical, biochemical and chemical engineering publications world-wide.

Published by the American Chemical Society, CA is divided into two parts. The abstracts themselves give short summaries of all papers containing chemical information, whereas the indices are the first port of call when you are looking for a particular piece of information. Each bound volume of CA is a collection of the weekly issues over a 6 month period (yearly prior to 1962) and has associated *Author, Chemical Substance, Formula* and *General Subject Indexes*. At 5-year intervals, the indices of ten volumes are combined in *Collective Indexes* (this interval was 10 years prior to 1957). Before 1972, the *Chemical Substance* and the *General Subject Indexes* were combined in the *Subject Index*.

Another useful work to be found in the reference section is the *Science Citation Index* (SCI), which is published every 2 months by the Institute for Scientific Information (ISI). Unlike CA, the literature coverage in SCI

Table 2.1 Primary literature sources

Acta Crystallographica C (crystal structure communications)
Advanced Materials
Angewandte Chemie, International Edition in English
Bulletin of the Chemical Society of Japan
Chemical and Engineering News
Chemische Berichte
Chemistry
Chemistry in Britain
Chemistry of Materials
Inorganic Chemistry
Inorganica Chimica Acta
J. Chem. Soc., Chemical Communications
J. Chem. Soc., Dalton Transactions
Journal of Materials Chemistry
Journal of Molecular Catalysis
Journal of Organometallic Chemistry
Journal of Organic Chemistry
Journal of Solid State Chemistry
Journal of the American Chemical Society
Mendeleev Communications
Organometallics
Polyhedron
Progress in Solid State Chemistry
Tetrahedron
Tetrahedron Letters
Zeitschrift für Anorganische und Allgemeine Chemie
Zeitschrift für Naturforschung Teil B

The following general scientific publications increasingly contain primary articles on chemistry:
Nature, New Scientist, Science, Scientific American.

---

is selective (about 3500 periodicals from across all science). In the printed form, it is actually a collection of three indices:

- The *Source Index* is an alphabetic list of names of authors (or other sources) who published during the period being reviewed, and it gives the title and journal details of the publications.
- The *Permuterm Index* utilizes keywords from titles of articles in the journals covered by SCI. For each keyword or combination of keywords are listed the names of authors who have published on them and hence, by using the *Source Index*, references for the original work can be obtained.
- The *Citation Index* is constructed so as to enable 'forward searching' from a particular publication. The index takes the form of a list of first-named authors and their papers which have been cited in any paper published during the period being reviewed. The importance of this feature is that, given one key reference in an area, you can build up a

picture of how work in that area has progressed since that publication (assuming that subsequent workers have indeed recognized and referenced the earlier work).

Summaries of original results appear in reviews and books and as collated data in reference texts and this tertiary literature provides another means of accessing the initial published work in the primary literature. The examples in Table 2.2 are likely to be useful starting places for searches.

An increasing amount of the chemical literature is now accessible electronically. Most chemists have access to computers in the office, laboratory or departmental network and also in the library. Nearly all important computer databases are searchable online, and online access to primary sources including full-text versions of journals from the RSC and ACS is increasing. Up-to-date information about online searching for chemical information is contained in the journals *Online* and *Database* (published by Online Inc.), the *Journal of Chemical Information and Computer Sciences* (published bimonthly by the ACS) and *Online Review* (published by Learned Information). The chemical information covered by the many databases presently offered by the several hundred vendors (hosts) can be divided into: bibliographic, full texts, structures, data about

Table 2.2 Tertiary literature sources

Accounts of Chemical Research
Advanced Inorganic Chemistry (F.A. Cotton and G. Wilkinson, 5th edn, Wiley)
Advances in Inorganic Chemistry (formerly *Advances in Inorganic Chemistry and Radiochemistry*)
Advances in Organometallic Chemistry
Chemical Reviews
Chemical Society Reviews
ChemTracts (Inorganic and Macromolecular sections)
Comprehensive Coordination Chemistry
Comprehensive Inorganic Chemistry
Comprehensive Organic Chemistry
Comprehensive Organometallic Chemistry
Coordination Chemistry Reviews
Dictionary of Inorganic Compounds
Dictionary of Organometallic Compounds
Encyclopaedia of Inorganic Chemistry
GMELIN Handbook of Inorganic and Organometallic Chemistry
Heterogeneous Chemistry Reviews
Inorganic Syntheses
Progress in Inorganic Chemistry
Progress in Solid State Chemistry
Russian Chemical Reviews
Specialist Periodic Reports
The Chemistry of Functional Groups (a series edited by S. Patai and published by Wiley)
Topics in Current Chemistry

chemical substances, chemical reactions. The vendor most likely to be used in universities is perhaps the Scientific and Technical Information Network (STN), since it offers several important databases at a reduced academic rate.

Chemical Abstracts is available online covering the literature from 1967 onwards and the *Science Citation Index* is available to academics *via* the Bath ISI Data Service (*BIDS*). STN provides searchable access to full-text journals published by the ACS (18 journals, 1982 onwards), RSC (ten journals, 1987 onwards), VCH (*Angew. Chem., Int. Ed. Eng.*) and Wiley (five polymer journals, 1987 onwards) in addition to the *Journal of the Association of Official Analytical Chemists*. Many of the developments in inorganic chemistry have been made possible by advances in X-ray crystallography, and it follows that the ability to search databases of crystal structures represents a powerful tool for the synthetic chemist. Important crystal structure databases that are available online are:

- The *Cambridge Structural Database* (compiled by the Cambridge Crystallographic Data Centre; this contains all organic and organometallic structures determined since 1935 and is also available for 'in-house' searching).
- The *Inorganic Crystal Structure Database* (produced by the Institute of Inorganic Chemistry of the University of Bonn, this contains the structures of over 24 000 inorganic compounds).
- The *NIST Crystal Data Identification File* (produced by the US National Institute for Standards and Technology, this contains over 100 000 structures).
- The NRCC Metals Crystallographic Data File, *CRYSTMET* (produced by the National Research Council of Canada, this contains structures and bibliographic details of more than 6000 metallic compounds).

The major producers of publicly available chemical substance databases for online searching are the Chemicals Abstract Service (CAS), Chemical Informations Systems Inc., the Gmelin Institute and the Beilstein Institute. STN and Questel offer structure searching (i.e. where atom connectivity is defined) of databases based on the CAS Chemical Registry System and this facility is similarly available for CIS. *Gmelin* and *Beilstein*, the major handbooks for inorganic and organic substances, respectively, represent the largest online sources of data in terms of numbers of substances. In addition to substance identification, these sources also provide preparative, physical and spectroscopic details. The *Dictionary of Organic Compounds* and *Dictionary of Organometallic Compounds* are also now available online together *via* the vendors Dialog and the Life Science Network. Mass, NMR and IR spectral data are also searchable online through several systems. Databases of organic reactions have been developed and, of these CAS React, REACCS and ORAC are most likely to be available. The Chemical Database Service (CDS) at Daresbury in the UK provides a service for the retrieval of chemical data, the graphical display of structures and the analysis of numerical data. The service is free to academics based at UK Higher Education Institutes and paid access can be arranged for industrial and overseas academic users.

It is clear that computerized methods of searching the chemical literature are evolving constantly and that they are being increasingly used by bench chemists as well as information specialists. In addition to the obvious advantages of speed and reliability, there are also some searches which cannot be carried out with the paper indexes: for example, searching the full text of major journals and reference works, using logic operators to combine searches, or searching for compounds containing a particular substructure. However, there are some drawbacks. The cost of hardware is significant and, even if your institution has all the necessary computer links, access to many of the databases must be paid for and the cost may be excessive for the group or departmental budget. A thorough understanding of the software is essential to make proper use of these facilities and this requires a significant investment in terms of time.

You should find out from your librarian, supervisor and colleagues what services are available and check whether any training is provided for their use, although users are usually given documentation. It is likely that you will make increasing use of these powerful tools as your research progresses.

2.2.2 Keeping up to date

During your self-allocated time in the library every week it will, of course, not be possible to read every journal associated with your interests from cover to cover. However, this is when moments of inspiration are most likely, and interesting revelations arising from chance observations while browsing may lead to a new approach to a problem or even a new avenue of research. It is, therefore, important that you develop the ability to scan contents pages, recognize tell-tale clues to the papers that might contain relevant material and then spot any valuable information without getting bogged down in the discussion and reasoning of the full text.

Significant new findings are very often published first as communications in journals such as *J. Chem. Soc., Chem. Commun.* and *Angew. Chem., Int. Ed. Eng.*, which are dedicated to this purpose (the latter also contains review articles) or in the appropriate section of other major journals mentioned in Table 2.1. It is fairly straightforward to pick out the important facts from this type of article. From a practical viewpoint, they don't usually contain full experimental details, but be sure to look at any footnotes – this is often where any description of procedures and spectroscopic characterization appears.

Full papers take a little more effort. You can normally get some flavour of the work by scanning the abstract, looking at any reaction schemes and figures (especially for structural formulae and crystal structure determinations) and by glancing through the experimental section. Important information about techniques, preparation and purification of starting materials, reagents and solvents is often given in the general, introductory paragraphs of the experimental section.

It is important that you make a note of anything that catches your eye immediately so always have a pen and paper with you. It is also a good idea to keep a record of which issues of each journal you have looked at to avoid any duplication of effort.

Current Contents, Physical, Chemical and Earth Sciences, published by ISI, contains a wide range of contents pages and is useful if your library doesn't take the particular journal in which you have an interest. You can then obtain copies of articles as inter-library loans. Since 1991, *Current Contents on Diskette with Abstracts* has been available (for a price) and enables some retrospective searching from a desktop computer.

2.2.3 Searching for specific information

Frequently you will need to search for specific types of information such as:

• physical and spectroscopic properties of a compound and how to synthesize it;
• examples of particular types of compounds or reactions;
• research already carried out in an area unfamiliar to you.

In this situation, it is important that you first define what you are looking for and then pick the brains of your supervisor and colleagues, who might have direct information or be able to help with your next decisions – which information sources to use and where to start looking. Keep detailed records of your searches in a separate hard-backed notebook. The records should contain all sources, references and notes about subject matter and should enable you to repeat a search at any time in the future. As such they are a valuable resource – they represent a lot of time, effort and expertise!

Physical properties of substances are often found in close-to-hand reference texts such as the *CRC Handbook of Chemistry and Physics* (the so-called 'Rubber Book'), older editions of which are usually to be found in laboratories, and in catalogues of suppliers of fine chemicals (Aldrich, BDH, Fluka etc.). General information can be obtained from *Advanced Inorganic Chemistry* (F. A. Cotton and G. Wilkinson, 5th edn, Wiley Interscience, 1988). Although usually regarded as an undergraduate textbook, the in-depth coverage of the elements (especially the transition metals), references to review articles and original papers and sections on organometallic and bioinorganic chemistry make this an excellent source of basic information. The 6th edition, presently being prepared, will update the coverage. For detailed information about specific compounds, however, you will probably need to use CA or *Gmelin*. If you are fortunate enough to have *Gmelin* in your library, it is not as difficult to use as it might first appear. The classification system allocates numbers to elements as listed in Table 2.3.

Note that system numbers are not identical with atomic numbers, as shown in Table 2.4. A compound will be listed under the constituent element with the highest system number. The annual 'Complete Catalogue' explains the *Gmelin* system and lists all volumes and supplements alphabetically according to element symbol and includes dates for literature coverage. This also contains an explanation of the ligand classification system used for organometallic and coordination compounds.

To find a particular compound in *Chemical Abstracts*, it is probably best to start with the *Formula Index* (and it is advisable to read the introductory pages first). Start with the most recent index and work backwards in time. Formulae are listed in order of increasing number of carbons, then hydrogens, and then increasing numbers of other elements

Table 2.3 *Gmelin* system numbers

No.	Element	No.	Element	No.	Element
1	Noble gases	25	Caesium	49	Niobium
2	Hydrogen	26	Beryllium	50	Tantalum
3	Oxygen	27	Magnesium	51	Protactinium
4	Nitrogen	28	Calcium	52	Chromium
5	Fluorine	29	Strontium	53	Molybdenum
6	Chlorine	30	Barium	54	Tungsten
7	Bromine	31	Radium	55	Uranium
8	Iodine	32	Zinc	56	Manganese
9	Sulphur	33	Cadmium	57	Nickel
10	Selenium	34	Mercury	58	Cobalt
11	Tellurium	35	Aluminium	59	Iron
12	Polonium	36	Gallium	60	Copper
13	Boron	37	Indium	61	Silver
14	Carbon	38	Thallium	62	Gold
15	Silicon	39	Lanthanides	63	Ruthenium
16	Phosphorus	40	Actinium	64	Rhodium
17	Arsenic	41	Titanium	65	Palladium
18	Antimony	42	Zirconium	66	Osmium
19	Bismuth	43	Hafnium	67	Iridium
20	Lithium	44	Thorium	68	Platinum
21	Sodium	45	Germanium	69	Technetium
22	Potassium	46	Tin	70	Rhenium
23	Ammonium	47	Lead	71	Transuranic elements
24	Rubidium	48	Vanadium		

H 2																	He 1
Li 20	Be 26											B 13	C 14	N 4	O 3	F 5	Ne 1
Na 21	Mg 27											Al 35	Si 15	P 16	S 9	Cl 6	Ar 1
K 22	Ca 28	Sc 39	Ti 41	V 48	Cr 52	Mn 56	Fe 59	Co 58	Ni 57	Cu 60	Zn 32	Ga 36	Ge 45	As 17	Se 10	Br 7	Kr 1
Rb 24	Sr 29	Y 39	Zr 42	Nb 49	Mo 53	Tc 69	Ru 63	Rh 64	Pd 65	Ag 61	Cd 33	In 37	Sn 46	Sb 18	Te 11	I 8	Xe 1
Cs 25	Ba 30	Ln 39	Hf 43	Ta 50	W 54	Re 70	Os 66	Ir 67	Pt 68	Au 62	Hg 34	Tl 38	Pb 47	Bi 19	Po 12	At 8a	Rn 1
Fr 25a	Ra 31	Ac 40	104 71														

NH4 23

Lanthanides 39	Ce	Pr	Nd	Pm	Sm	Eu	Gd	Tb	Dy	Ho	Er	Tm	Yb	Lu

Actinides	Th 44	Pa 51	U 55	Np 71	Pu 71	Am 71	Cm 71	Bk 71	Cf 71	Es 71	Fm 71	Md 71	No 71	Lr 71

Table 2.4 Periodic table showing *Gmelin* system numbers.

in alphabetical order. From the list of names under each formula, try to identify the compound you want and use that name to look in the *Chemical Substance Index*. The keywords given there for each abstract number make it easier to identify abstracts containing information of interest.

For general areas and classes of compounds, it is probably best to start with the tertiary sources listed in Table 2.2, but always check the period of literature coverage. In rapidly developing fields, articles can quite quickly become outdated. The same applies to monographs. Reviews are somewhat more flexible and can be published more quickly. If you have found an older review or monograph, later work can be traced by using the *Science Citation Index* as described above.

When carrying out more exhaustive searches, once you have identified key papers follow them up using *SCI*. Take full advantage of any computerized searching facilities, especially those with structural input.

2.2.4   Organizing the information

At some stage all the information you have obtained must be organized systematically, even those articles you have photocopied for future reference. This is usually done using a card index system, although there are now bibliographic software packages available which often allow direct downloading of references from online searches. These records are effectively an extension to your memory and you will have to develop a method of classification which suits you best, but remember that as time goes by it may be increasingly difficult to change a particular system once you have started to use it.

2.3 Keeping records

The importance of recording details of literature searches and of developing a systematic method of storing the information obtained has been emphasized above. Exactly the same principles apply to experimental work. Accurate details of every experiment should be recorded in a hard-backed notebook and they should be sufficient to enable you or someone else to repeat the work. Products should be labelled clearly to identify them with a particular experiment, and stored so that they can easily be found in the future. Associated spectra and other data should be clearly labelled to identify them with particular products of any experiment and then filed in a systematic fashion to allow ready retrieval. This level of discipline is essential for successful synthetic work and anything less constitutes a sloppy approach. Different groups develop their own styles for recording their work but several important features should be common to them all. In the following sections, the essential features are illustrated with reference to one particular format for keeping records (which happens to be more or less the one I was shown at the start of my PhD!).

2.3.1 The lab notebook

Your time in any one research group is unlikely to be more than 3–5 years, and when you move on your legacy is a set of lab notebooks which then become part of the group archives. Other workers will refer to them in their own projects, so it is important (for your own reputation, if nothing else) that you use your notebooks properly. The lab notebook serves several purposes:

- To provide an accurate record of every experiment you carry out. In this regard it is effectively a legal document which establishes what you did and when you did it, and also allows procedures for any experiment to be repeated or modified.
- To enable the identification of your samples.
- To enable the location of spectroscopic and other data from specific experiments.

No.	Experiment	Anal.	IR	^{13}C	^{1}H	Mass	Xtal
RJE 1	Prepn. of $WOCl_4$	✓	✓	—	—	—	—
RJE 2	Prepn. of $WO(OMe)_4$	✓	✓	VT	✓	✓	
RJE 3	Att'd prepn. of $WO(OCH_2CH_2OMe)_4$	✓	✓	✓			

Figure 2.2   An example of a contents page.

- To be the central reference source for reports and publications based on the work and, of course, for your thesis.
- To help you keep track of your work. Following the progress of several concurrent experiments can be difficult if you don't record each step and label products accordingly.

When starting a new lab notebook always leave the first few pages blank for a table of contents. Each time you start an experiment, enter the experiment number and title in the contents pages. Leave some space to indicate whether elemental and spectroscopic analyses, etc. have been obtained so that you can see at a glance which ones are still required. A suggested layout for a contents page is shown in Figure 2.2.

Use two facing pages for each experiment, with the main write-up on the right-hand page. Put the number of the experiment (your initials plus a number) on the outside corner of each page. The title can then go at the top of the right-hand page with the date (very important) on the inside edge. On the left-hand page (Figure 2.3) write down any key references and relevant equations and list the starting materials and expected products together with their formula weights. As you weigh out the reagents, the weighings of flasks or other containers should also be entered on this page in addition to the amounts used. Calculations of percentage elemental composition are best written out in full so that they can easily be checked for mistakes. Microanalysis results can then be entered next to the calculated values. If the apparatus is special in any way, a rough sketch will help if the experiment has to be repeated. Sketches of NMR spectra and results of crystal structure determinations are also useful.

On the right-hand page (Figure 2.4), your description of the procedure should be succinct and contain all relevant information. It should be written directly into your lab notebook as you carry out the experiment and not copied later from notes on scraps of paper. Note down all observations such as changes in colour and temperature, gas evolution, precipitate formation, etc. You don't have to write in sentences, and the use of abbreviations will help you save time and space, e.g. R for 'room

$$WOCl_4 + 4\ MeOH \rightarrow WO(OMe)_4 + 4\ HCl$$

Ref: *J. Chem. Soc., Dalton Trans.*, 1992, 1431

$WOCl_4$	340.63	2.54 g	7.46 mmol		
MeOH	32.04	0.96 g	29.83 mmol	$\rho = 0.791\ g/cm^3$	$1.21\ cm^3$
$WO(OMe)_4$	323.99	2.46 g	7.46 mmol	theoret	

$C_4H_{12}O_5W$		% calc	% found
C_4	48.044	14.83	14.78
H_{12}	12.096	3.73	3.82
O_5	80.000	24.69	
W	183.850	56.75	
Σ	323.99	100.00	

flask + $WOCl_4$ 81.07 g
flask................ 78.53 g
$WOCl_4$ 2.54 g

flask + $WO(OMe)_4$ 100.47 g
flask....................... 98.1 g
$WO(OMe)_4$ 2.37 g

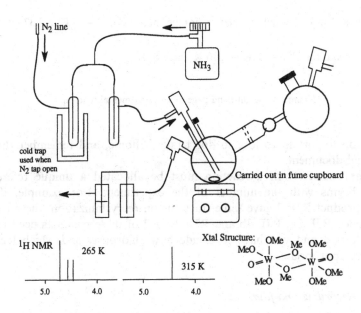

Carried out in fume cupboard

Figure 2.3 Left-hand page of an experimental write-up.

temperature', O/N for 'overnight'. A reaction which involves an extended work-up will take up a lot of space as you describe all the separations and recrystallizations, but try and keep it within the two pages of your lab book – even if it means reducing the size of your handwriting. Which reminds me – handwriting. This must be legible, otherwise it defeats the whole object of the exercise. Keep your lab notebook in the lab, it is

Preparation of WO(OMe)$_4$

WOCl$_4$ (2.54 g, 7.46 mmol) in THF (80 cm^3). Stirred → orange soln.
Cooled (liq. N$_2$/acetone).
10.00 MeOH (1.4 cm^3, 34.56 mmol) syringed in + stirring.
10.20 bath removed, allowed to warm → R°.
10.50 soln. very pale yellow.
N$_2$ bubbled thro' soln. 30 min then NH$_3$ for 30 min → white ppt.
Filtered, solid washed + THF (20 cm^3), solution stripped → off-white solid.

2/10/99
Dissolved in THF (10 cm^3) + hexane (15 cm^3) with heating → freezer.

3/10/99
Colourless Xtals. Filtered, washed + hexane, pumped dry → RJE 2 2.37 g 98%

Sample prepd. for ^{183}W-NMR @ 370 K and 340 K.

Figure 2.4   Right-hand page of an experimental write-up.

much too important to lose so don't take it home, and remember that it is a legal document.

Each product from a reaction must be allocated a unique reference which begins with the number of the experiment. For example, if an initial product RJE 3 gave three fractions on recrystallization, these might be labelled RJE 3a, RJE 3b and RJE 3c. All of these products need to be characterized and you have to decide how rigorously and which techniques are appropriate.

2.3.2   Recording and filing data

The degree of characterization necessary for a given compound will depend upon the nature of the product, but for a pure, previously uncharacterized organometallic or metalorganic substance, the normal data set will be elemental analysis, IR, ^{1}H- and ^{13}C-NMR (with other nuclei as necessary) together with UV/Vis, mass spectrometry (EI, CI, FAB or electrospray), cyclic voltammetry and/or an X-ray crystal structure determination where appropriate.

Ensure that the correct reference from your lab notebook is entered for each compound on the standard submission forms normally required by microanalytical and spectroscopic services and when you record your own spectra or other data write the reference number on them immediately. File all of these records numerically. For each compound, I prefer to keep the combined data along with photocopies of key references in one file for easy access, and list the contents on the outside. Together with the details in the contents pages of your lab notebook, this makes it easier to spot any missing information and helps to ensure the full characterization of key compounds.

2.4 Safety

The chemical laboratory can be a hazardous place to work, but this does not necessarily mean that it is dangerous. A well-organized and well-managed lab is likely to be safe and pose minimal risks to health. Every substance or operation has an associated hazard – a potential to cause harm. The likelihood that harm will actually be caused is the risk associated with any process. It follows that to work safely you must:

- identify hazards associated with all substances and procedures you intend to use;
- minimize the probability that any harm will be caused by these combined factors.

What follows is a summary of the important points of laboratory safety [4]. No list of advice can cover all eventualities, and you will be required to read the department's own safety booklet when you start so, in addition to emphasizing the basics, this section is intended to provide guidance on the development of safe working practices. Useful information is tabulated in Appendix A.

2.4.1 Legal aspects

The introduction of the Control of Substances Hazardous to Health (COSHH) Regulations (1988) and the Management of Health and Safety at Work Regulations (1994) in the UK and the Toxic Substance Control Act (TSCA) in the USA has imposed new responsibilities upon the management of chemical companies and universities. Where safety checks might previously have been taken for granted, extensive documentation of risk assessments and any action taken as a result is now required. The result is an increasing bureaucracy in preparative chemistry laboratories but, while formalized procedures are somewhat irksome, they do reinforce good chemical practice and provide documentary evidence that proper consideration has been given to any hazards and risks associated with the work.

2.4.2   Responsibilities

Another result of the legislation is that most institutions will have set up a well-defined chain of responsibility for safety matters from the top downwards. However, this does not affect the bottom line – **you** are responsible for your own safety and that of your colleagues in the lab, since this is determined mainly by **your** working practices.

2.4.3   General behaviour and awareness

Common-sense should prevail while you are in the lab, which means adhering to some straightforward codes of conduct.

- Always wear safety spectacles.
- Don't eat, drink or smoke in the lab and don't work after drinking alcohol.
- Work carefully, don't rush anything and don't become complacent.
- Avoid working alone, someone within earshot must be aware of your presence.

In addition, you should always be prepared for any accidents or emergencies. Every so often, ask yourself, 'What would I do if something happened now?' If you are uncertain of the answer, you should not be working in the lab.

- Know the location of eye washes and/or safety shower.
- Know the emergency exit routes, the evacuation procedure and the alarm sound.
- Know the location of fire-fighting equipment and how to use it.
- Know how to summon help (first-aid, ambulance, fire service).

2.4.4   Risk assessment

The key to safe working is a knowledge of all the chemical and physical hazards that might be encountered during an experiment so that you can make an assessment of the risk before you start. You will then be in a position to take the necessary precautions to minimize the risk.

Hazards which present a risk of physical damage to your person are usually fairly obvious. The most common injuries by far are cuts from broken glass, often caused by poor technique or sloppy house-keeping. Put broken glass in bins clearly labelled 'GLASS ONLY'. Nothing else should go into these bins. The correct way to attach flexible hoses to glassware is described in chapter 3 in the section on Schlenk techniques. Evacuated or pressurized apparatus, electrical equipment, cryogenic liquids and sources of non-ionizing radiation (lasers, UV, IR and microwave) must all be considered and local rules may exist for their use.

This also applies to lifting and moving equipment (Manual Handling Operations Regulations) and extended use of computers (Display Screen Equipment Regulations).

You should treat all chemicals as potentially dangerous and handle them accordingly, but for the risk assessments required by the COSHH regulations [5] you need to check and note down the known hazards of every compound you encounter. This is not as arduous as it sounds and should in any case be instinctive. Suppliers are required to provide such details when they sell a chemical, but the information often arrives some time after the compound. Collections of Hazard Data Sheets or Materials Safety Data Sheets are available from companies such as Aldrich and BDH and other compilations are available [6,7]. There is also a wide range of online and CD-ROM databases and an excellent survey is given in ref. [1].

Once you have the information, hazards can be graded [5] from extreme to low according to Table 2.5. You then have to estimate the likely exposure to the various substances, taking into account the quantity and physical attributes of each (e.g. volatility, dustiness and concentration) and the nature of the procedure (likelihood of airborne material or exposure *via* skin contact, inhalation, ingestion, inoculation; frequency and duration of activity). By combining the hazard and exposure estimates the procedure can be classified as high, medium or low risk and you can then decide upon a suitable degree of containment. Details of this system of risk assessment are given in ref.[5]. Records of risk assessments must be kept in all but the most trivial cases where action is simple and obvious. When preparing the assessment consult your supervisor about unfamiliar aspects (especially techniques you haven't used before). The record must then be read, signed and dated by your supervisor or another suitable competent person and must be kept

Table 2.5 Guidelines for determining a chemical hazard category [5]

Extreme	Where exceptional toxicity is established or suspected (e.g. carcinogens)
High	Where toxicity exceeds that of the medium category, except for substances in the extreme category
Medium	'Harmful' or 'Irritant' substances according to CPL[a] classification
Low	Where substance does not meet CPL[a] criteria for classification as 'Harmful' or 'Irritant'

[a]CPL, Classification, Packaging and Labelling Regulations 1984, superseded by: CHIP, Chemicals (Hazard Information and Packing) Regulations, 1993.

close to where the work is being carried out so that all the information is available in the event of an accident.

In most cases, the containment afforded by the use of a combination of Schlenk and dry box techniques is sufficient for most substances although those in the 'extreme' hazard classification may require extra precautions. Specific safety warnings are given where appropriate throughout the book and further information is given in Appendix A.

References

1. Bottle, R.T. and Rowland, J.F.B. (eds) (1993) *Information Sources in Chemistry*, 4th edn, Bowker and Saur, London.
2. Wolman, Y. (1988) *Chemical Information, a Practical Guide to Utilisation*, 2nd edn, Wiley, Chichester.
3. Maizell, R.E. (1987) *How to Find Chemical Information*, 2nd edn, Wiley, Chichester.
4. *Safe Practices in Chemical Laboratories*, (1989) The Royal Society of Chemistry, London.
5. *COSHH in Laboratories*, (1989) The Royal Society of Chemistry, London.
6. Luxon, S.G. (1992) *Hazards in the Chemical Laboratory*, 5th edn, The Royal Society of Chemistry, Cambridge.
7. Urben, P.G., Pitt, M.J. and Battle, L.A. (1995) *Bretherick's Handbook of Reactive Chemical Hazards*, 5th edn, Butterworth-Heinemann, Oxford.

3 Bench-top techniques

3.1 Introduction to inert atmosphere techniques

Many of the compounds encountered in metalorganic chemistry are reactive towards moisture and/or oxygen and must be isolated from the lab atmosphere and handled in a controlled environment. The ultimate controlled environment is, of course, a vacuum, and the original chemical vacuum line was developed by Alfred Stock in the early 1900s to handle volatile, pyrophoric boron hydrides. Over the years, with the availability of new materials and more modern equipment, the original design has evolved and more flexible systems which include an inert gas supply (usually nitrogen or argon) can now be constructed readily. Workers today usually use a combination of techniques, the choice being determined mainly by the properties of the compounds being handled, and also to some extent by the past experience and preferences of the personnel involved.

Vacuum/inert gas manifold systems are fairly simple in design, straightforward to use and are easily accommodated on the bench top or inside a fume cupboard. When used in conjunction with specially designed glassware, this is the usual method of choice for routine exclusion of air and it provides a high degree of flexibility. This chapter describes various aspects of this approach. When a more rigorous exclusion of air is required, or a reaction involves more volatile compounds, it may be necessary to use a high vacuum line as described in chapter 5. This apparatus is perhaps more difficult to use than the bench-top manifold, but it does allow the quantitative manipulation of condensable and non-condensable gases. Inert-atmosphere glove boxes (often called dry boxes, chapter 4) are used for conventional manipulations of air-sensitive materials, and although it is possible to carry out reactions within a dry box, they are more often used for the weighing and transference of solids or liquids and for the preparation of samples for IR and NMR spectroscopic investigation while the synthetic work is done on a manifold or vacuum line.

Two publications [1,2] provide excellent discussions of most aspects of this type of work, and I recommend that they be consulted for more detailed descriptions of some of the techniques covered in the following chapters, especially if you intend to design and build new apparatus. Several older texts, some of which contain synthetic details for selected compounds, also give accounts of the techniques and apparatus which can be used to handle air-sensitive materials [3–6].

3.2   The inert-gas/vacuum double manifold

3.2.1   Design

When you start in the lab, it is likely that you will inherit an inert-gas/ vacuum manifold along with your bench space from a previous research student. If not, and if you have ready access to glassblowers, then you will need to produce a design and have one made; otherwise commercial versions are available from various suppliers. The basic design shown in Figure 3.1 allows the atmosphere inside a piece of apparatus to be readily switched between vacuum and an inert gas simply by turning a two-way tap, and normally four or five taps are incorporated so that several items can be connected to the line at the same time.

The ends of the two glass tubes have joints for connection to the inert-gas supply and to a vacuum pump *via* a low temperature trap. The incorporation of two joints on the vacuum tube enables the pump to be positioned to either side of the manifold on the bench top. To make cleaning easier (something you will appreciate later) the joints should be large enough for a long, wire-handled 'test-tube' brush to be inserted and removed. The convenient combination of female B24 ground-glass joints and tubing of approximately 25 mm o.d. provides a sturdy, durable construction for clamping to the framework, although substitution of ball and socket joints for the B24 vacuum connections makes life easier when

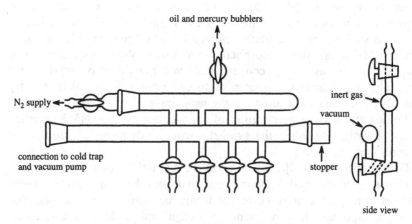

Figure 3.1   Vacuum/inert-gas manifold.

it comes to aligning and clamping the trap to the manifold. Double oblique, 4-mm bore high-vacuum taps are connected to the glass tubes as shown in the side view of the manifold in Figure 3.1. The spacing between the taps can be tailored to the space available at the bench, but I've found 120–150 mm to be adequate while maintaining the compact nature of the design. Lengths of flexible rubber or plastic hose (Tygon or Portex PVC tubing is commonly used) attached to the taps are used to connect apparatus to the manifold. A wall thickness of at least 3 mm is desirable so that the tubing will not collapse under vacuum, and these hoses should only be of sufficient length to reach the bench top; long trailing tubes are unwieldy and can knock pieces of glassware off the bench. Bear in mind that rubber and plastic tubing is slightly permeable to oxygen and moisture and can also absorb solvent vapours.

The second tap on the inert gas side of the manifold is connected to a mercury manometer/bubbler in conjunction with an oil bubbler and non-return valve (a suitable inexpensive plastic valve is available from Aldrich, which should be mounted vertically to ensure that the valve closes when the gas flow stops). With this arrangement, nitrogen bubbles out through the oil bubbler when the pressure in the nitrogen line is above that set by the level of oil in the bubbler and, when the pressure is reduced, the non-return valve closes and the mercury rises in the manometer tube. This minimizes the entrainment of mercury vapour in the exhaust gases from the line, which must be led to the fume cupboard.

The low temperature trap connected between the manifold and the vacuum pump is shown in Figure 3.2. This prevents volatiles from

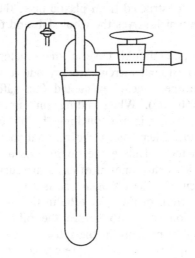

Figure 3.2   Cold trap to protect a vacuum pump.

entering the pump, and the high vacuum isolation tap between the trap and the manifold should have a large bore so as not to reduce the pumping efficiency. The small tap allows the trap to be vented to air once the pump has been switched off and the liquid nitrogen Dewars removed (to prevent condensation of liquid oxygen), so that pressure does not build up while the trap warms to room temperature.

3.2.2 Setting up the manifold

If you have a new manifold or one which has just been cleaned, the various components must be assembled and clamped to the framework with care to avoid placing the glassware under any strain. The manifold should first be clamped with the taps at a comfortable working height. Adjustment of the vertical position and the distance from the framework is easier if clamps are attached to horizontal rods, which in this case is not a problem provided two rods are used for support. If the clamps are arranged with their movable jaws uppermost, the manifold can be supported on the fixed lower jaws while final adjustments are made before the clamps are tightened. This also means that these clamps need not be moved when the system is dismantled for cleaning; after disconnecting the trap and bubblers, the manifold can be simply lifted out after opening the clamp jaws. The lower jaws remain in the correct positions for reassembly. The arrangement of the assembled manifold, trap and the oil and mercury bubblers is shown in Figure 3.3a.

Note: The exhausts from the bubblers and from the vacuum pump must be led to the fume cupboard, and the neatest solution is to build an exhaust manifold with an internal diameter of about 20 mm on each bench. Conveniently constructed from plastic pipe, these can be linked to form an exhaust line which serves the whole lab and terminates in a fume cupboard.

The mercury bubbler should have a manometer tube longer than 760 mm and contain enough mercury to provide a seal from the atmosphere when the nitrogen line is evacuated (i.e. sufficient to provide a column of about 760 mm). When setting this up, first add sufficient mercury so that the surface is just touching the bottom of the inlet tube from the nitrogen line. Then, using the internal diameter of the tubing used in the manometer, calculate the volume of a 760 mm column of mercury and add at least this amount of extra mercury to the reservoir.

The working pressure in the nitrogen line is set by the height of liquid paraffin above the bottom of the inlet tube in the oil bubbler (about 60–100 mm). Incorporation of a tap before the oil bubbler provides for occasions when a higher pressure in the nitrogen line (determined by the height of mercury in the manometer reservoir) is required for short periods.

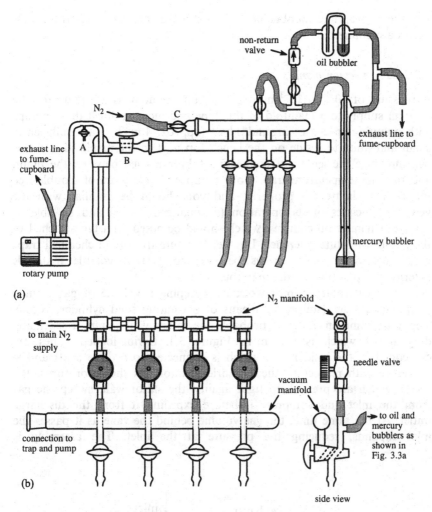

Figure 3.3 (a) Complete manifold system. (b) Alternative inert-gas supply arrangement providing independent pressure control.

Alternatively, by incorporating a mercury and an oil bubbler at each position, a choice of two pressures (in the same manner as described above) is provided to each two-way tap on the line. This arrangement is shown in Figure 3.3b, and enables the gas flow to each two-way tap to be controlled independently by a needle valve (Figure 6.8a). This means that the pressure differential required for cannula transfer or filtration (section 3.4) can be achieved by having the receiver at oil-bubbler pressure while the pressure above the liquid is that in the mercury bubbler. The nitrogen supply line in this case is constructed readily from 6 mm copper tubing and the appropriate number of T- and elbow-joints, to which the needle valves are connected.

3.2.3   The vacuum pump

A pressure of 10^{-1} to 10^{-3} mmHg (medium vacuum) is adequate for the vacuum supply to a manifold of this type and modern direct drive pumps with pumping speeds of around $1 \, 1 \, s^{-1}$ (approx. $4 \, m^3 \, h^{-1}$) are sufficiently small and quiet to reside on the bench top next to the manifold and this will minimize the length of connecting tubing needed between the pump and the low-temperature trap. Your pump will give years of trouble-free service if it is treated properly, and you should be familiar with the general principles of its operation (summarized below) and be able to carry out regular oil changes, which should be noted on a tag attached to the pump for future reference. Low-temperature traps (and chemical traps where necessary) must **always** be used to prevent volatile materials entering the pump when the manifold is in use.

Mechanical rotary pumps work by sweeping a volume of gas from a pumping cylinder (stator) by means of a rotating solid cylinder (rotor). The most common design is probably the rotary vane pump, a schematic diagram of which is shown in Figure 3.4. Spring-loaded vanes are embedded in the solid rotor, which is connected to a drive shaft and is off-centre with respect to the cylindrical stator. As the rotor turns, the vanes, which are pressed up tight against the stator wall, sweep the gas from the inlet and compress it before expelling it from the discharge outlet. At the same time, the gas volume behind the vane as it passes the inlet expands, reducing the pressure at the inlet. The assembly is

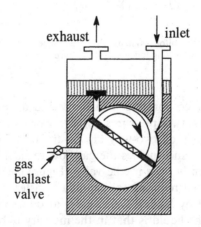

Figure 3.4 Rotary-vane vacuum pump.

immersed in oil in the pump casing; the oil acting as a seal and as a lubricant. In two-stage pumps, lower ultimate pressures are obtained (about 10^{-3} mmHg) by connecting two of these units in series, the outlet of the first being connected to the inlet of the second stage. In older models, power is transferred from the motor to the drive shaft by a belt, but direct drive *via* a semi-flexible coupling is now more common.

One feature often overlooked by research students is the gas ballast valve. When this one-way valve is open it allows a small amount of air into the pump and aids in the removal of condensable vapours from the oil, although the ultimate pressure is slightly increased as a result. When the manifold is operated correctly, the gas ballast should never be necessary but if you suspect that condensable materials have escaped the low-temperature trap and entered the oil, running the pump with the gas ballast valve open for a few hours could rectify the situation.

3.2.4   The inert gas supply

The inert gas (normally nitrogen or argon) is either fed directly to the manifold or it is first passed through drying and/or deoxygenation columns (see below). In some departments the boil-off from a main liquid nitrogen tank is used to provide an in-house nitrogen gas supply; otherwise the gas will be supplied from cylinders. Rather than use one cylinder for each manifold, which takes up a lot of floor space and entails a large expenditure on cylinder rental, a distribution line fed from a single cylinder can be constructed from copper tubing with needle valves to supply each bench position.

3.2.5   Purification columns

If there is reason to suspect the purity of the inert gas supply, water content can be reduced by passage through 3A or 4A molecular sieves, and efficient oxygen scavengers can either be purchased or prepared (Appendix B). These materials are normally packed into a glass column of the type shown in Figure 3.5, and a heating element wound onto the column (which is then enclosed in a protective cylindrical glass jacket) enables regeneration to be carried out *in situ;* otherwise heating tapes can be used. A highly active form of supported copper metal is commercially available in pellet form as BTS Catalyst, and is a very convenient oxygen scavenger, while the absorbents described in Appendix B have the advantage of being self-indicating. A supported, highly active form of MnO changes from green to brown upon oxidation, while silica-supported CrO changes from blue to green/brown. All of the above oxygen scavengers can be regenerated by treatment with hydrogen (as a nitrogen/hydrogen mixture) at elevated temperatures although, with

Figure 3.5 Inert-gas purification column.

careful use, this need only be done about once a year. Note that water is generated during this process and it is best to evacuate the heated, regenerated column to remove the last traces of moisture.

3.2.6 Initial setup

Once the line has been assembled and connected to the vacuum pump, an inert gas supply (which from here on will be assumed to be N_2) and the oil and mercury bubblers as in Figure 3.3, the following procedure can be used to check the system for leaks and purge it with N_2 prior to use.

- Grease the ground joint and connect the cold trap (which should be clean and dry). Secure the trap with springs or a rubber band looped over a nearby clamp to prevent it from falling onto the bench top when it is not under vacuum.
- Close the bleed tap (A) on the vacuum supply and also the main tap (B)

and ensure that all the double-oblique taps on the manifold are closed (i.e. with the key handles horizontal).

- Turn on the vacuum pump.
- Place a Dewar vacuum flask around the cold trap and fill it with liquid nitrogen (slowly at first while the Dewar flask cools, and then increase the level of nitrogen to just below the greased joint).
- Open tap (B) to evacuate the vacuum line.
- Connect two of the hoses together by a short piece of glass tubing. Open the tap above one of these hoses to vacuum.
- Ensure that the tap to the nitrogen supply (C) is closed and then **slowly and carefully** open the tap above the other hose to the nitrogen line. This will evacuate the nitrogen line (the non-return valve should close and prevent suck-back of oil) and the mercury will rise to about 760 mmHg in the bubbler, which is now acting as a manometer and can be used to check for leaks.
  Note: if you have insufficient mercury in the bubbler, air will be pulled through it and force mercury into the nitrogen line.
- Close the main tap (B) and observe the mercury level. Any decrease in height of the mercury column indicates a leak in the combined nitrogen and vacuum lines. Similarly, with the manifold taps closed, the evacuated nitrogen line can be isolated and leak tested. A Tesla coil (section 5.2.3) may also be used to detect leaks in the evacuated glassware.
- When you are happy that the system is leak-free, open tap (B) and the manifold taps to the two linked hoses and pump out the whole manifold for several minutes.
- Close the manifold tap which is open to the nitrogen line and slowly open tap (C) and then any taps on the purification columns. The main valve (from the house supply line or on the cylinder) can then be opened to admit nitrogen. Control the rate at which nitrogen enters the line so that the mercury level drops smoothly until gas bubbles out through the oil bubbler.

After repeating this purging process several times (closing tap (C) each time before evacuation) the line is ready for use.

3.3 Schlenk techniques

3.3.1 Principles involved

When a flask is attached to the manifold and evacuated, the amount of atmospheric gases remaining in the flask is greatly reduced to a fraction (F) of its original value. If the flask is then filled with pure inert gas and the evacuation process repeated, the fraction remaining will be F^2

assuming that there are no leaks in the system. For example, if the pressure in the vacuum line is 10^{-2} mmHg, $F = 10^{-2}/760$ i.e. 1.3×10^{-5}, after n pump/purge cycles the fraction remaining is $(1.3 \times 10^{-5})^n$. Theoretically then, this technique enables good exclusion of oxygen and moisture without the need for a high vacuum, but its success is very much dependent upon the skill of the operator.

Once a flask has been purged in this manner, the stopper can be removed provided that a sufficient flow of inert gas, in through the side-arm and out through the neck, has been established.

3.3.2   Reaction flasks

A flask with a side-arm for use with an inert gas was described by Walter Schlenk in 1913, and this formed the basis for the wide variety of modern designs. The standard requirements of a joint to connect to other items of glassware and provide access to the contents, and a tap for connection to the manifold, leave the actual design very much to the preferences of the individual, and some typical examples are shown in Figure 3.6. The 'tube' and round-bottomed styles can be made easily in the glass workshop if you provide the glassblower with your preferred joints and taps, and a range of sizes from 10 to 500 cm^3 will be necessary for routine work. For the larger sizes (> 200 cm^3) round-bottomed flasks are best. It is useful to make the neck below the joint long enough (about 40 mm) to allow the flask to be clamped at this position.

3.3.3   Ground-glass joints and taps

A variety of different types of joints and taps is now available but, in general, the choice is between those that require grease and those that do not. The least expensive option is a combination of tapered ground-glass joints and interchangeable ground-glass taps with solid keys. Convenient joint sizes are B24 and B19, while B14 joints can be used for smaller items. The bore size of the taps should be 3 or 4 mm to allow a sufficient flow of gas into the flask. High-vacuum ground-glass taps are less prone to 'freezing' than cheaper solid taps, although they are considerably more expensive and the keys are not interchangeable (the key and barrel of this type of tap are usually numbered, but it can be difficult to find matching pairs once the apparatus has been dismantled and cleaned).

3.3.4   Greasing ground glass

Rigid joints in vacuum systems which are expected to remain in place for some time can be sealed with wax. This is supplied in stick form, and both parts of the joint must be heated with a hot air gun before applying

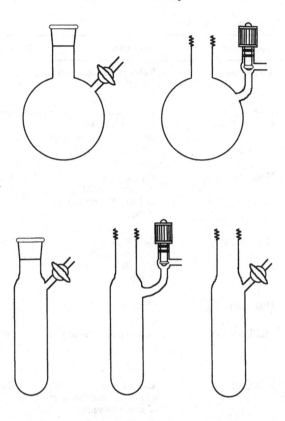

Figure 3.6 Round-bottomed and tube designs for Schlenk flasks incorporating ground-glass or screw joints and greased or greaseless taps.

the wax to the male joint. Both halves can then be pushed together, usually with further warming. Gentle heating to above the melting point is again required for disassembly.

For ground-glass surfaces to slide freely over each other they must be lubricated, and this is especially important when the apparatus is to be evacuated. On joints that are not intended to be permanent (stoppers, etc.), it is usual to use silicone grease, whereas an Apiezon grease which remains soft to lower temperatures is better for taps, so that they don't become too difficult to turn in winter. Some greases and their properties are listed in Table 3.1. A common problem with ground-glass joints and taps is the leaching of lubrication grease by solvents (although fluorinated greases are less soluble), causing surfaces to stick together or the introduction of impurities that are not always readily separated from reaction

Table 3.1 Vacuum greases

Type	Application	Properties	Temp. range
Apiezon H	Vac. joints	Rubbery, doesn't melt up to 250°C	−15°C to ambient
Apiezon L	High-vac. joints	Very low vapour pressure: 10^{-5} mm at 300°C 10^{-11}–10^{-10} mm at room temp.	Ambient to 30°C
Apiezon M	High-vac. joints	Vapour pressure: 10^{-3} mm at 200°C 10^{-8} to 10^{-7} mm at room temp.	Ambient to 30°C
Apiezon N	High-vac. stopcocks	Vapour pressure: 10^{-3} mm at 200°C 10^{-9}–10^{-8} mm at room temp.	Ambient to 30°C
Apiezon T	High-temp. vac. joints	Melts at 125°C	Ambient to 110°C
Dow Corning high-vacuum silicone grease	Multipurpose	Low vapour pressure	−40°C to 260°C
Krytox 240 AC (fluorinated grease)	Multipurpose	Stable towards oxidation and heat, insoluble in most common solvents	−30°C to 550°C
Krytox LVP	High-vac. systems	Similar to above	Similar to above

products. It is, therefore, important to use sufficient but not too much grease on joints and taps, so apply the grease only to the middle two-thirds of the length of the joint or tap key. Each person in the lab can have a small (2 or 5 cm^3) plastic syringe (without a needle) filled with grease from the main tube, enabling joints to be greased directly from the syringe. Alternatively, use a small piece of wooden or plastic dowel (like the type used in cotton buds). Try to avoid using a finger – this is a good way of applying grease where it isn't needed (flasks, clothing, etc.). Keep the applicator clean to prevent the introduction of dirt which will scratch surfaces and impair the seal, and always replace the cap on the tube. Use sufficient grease to give a clear, continuous film between the surfaces when one of them is rotated. 'Streaks' in the grease can often be removed by warming with a hot air gun while rotating the joint/tap. Should the streaks persist, the surfaces are dirty or ill-matched and you will have to clean the parts with a chlorinated solvent and start again. In extreme cases, the tap can be re-ground with successively finer grades of abrasive paste to obtain matching surfaces. This is perfectly within the capabilities of a research student, but it is advisable to check with your glassblower before commencing. If you need to put glassware down on the bench top during greasing and assembly, place it on a clean piece of paper towel to keep dust off the joints (a tap key should be inserted directly into the barrel after greasing). Occasionally, grease will react with the chemicals used or generated in a reaction, resulting in a frozen joint or tap which at best is a nuisance, but which can also be hazardous if pyrophoric substances are involved. Seized stopcocks can be freed by heating the barrel with a medium-hot flame from a gas torch. Wear oven gloves and apply even heat for about 15 s with a bushy flame before twisting and pulling out the key. Beware of flammable vapours which might ignite.

It is always best to remove all traces of grease from glassware before taking apparatus to the glassblower for repair. Although Apiezon greases are hydrocarbons and will burn off in the glassblower's oven, silicone greases will leave deposits of silica on joints and taps, which could render them useless.

3.3.5 Greaseless taps and joints

Where grease is likely to be leached by solvents or attacked by chemicals, it is often best to use greaseless joints and taps. Several types of tap based on the movement of a glass or plastic (Teflon or Kel-F) stem within a glass barrel are commercially available, and Figure 3.7a shows a typical example. The seal formed between the glass seat and a Teflon stem eliminates problems due to chemical attack and leaching of grease. In a design supplied by J. Young (UK), O-rings on the stem are protected from chemical attack by a sheath of PTFE which is an integral part of the stem. If kept clean and used properly, these taps perform well down to pressures $< 10^{-4}$ mmHg, although extremes of temperature should be avoided. Needle-valve versions of this type of tap are available when a controlled flow of gas or liquid is required, as opposed to a choice between on or off. Due to the higher thermal expansion coefficient of Teflon compared with glass, the seal will be lost when the valve is cooled unless it is re-tightened. Be careful though, these stopcocks are likely to break when overtightened. Always keep an eye on the compression of the Teflon O-ring or the tip of the stem against the glass seat.

The greaseless glass joints that are now available usually incorporate O-rings. A basic type is shown in Figure 3.7b, and this has the advantage that both parts of the joint are identical, thereby increasing apparatus flexibility. The two halves of the joint are held together by a compression clamp. In a modification of this design available from Kontes Glass

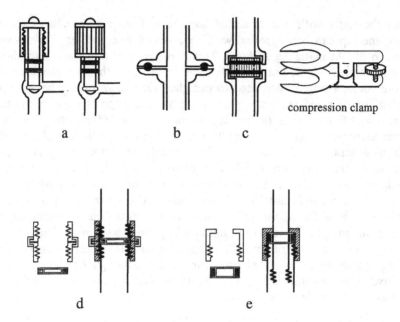

compression clamp

a   b   c

d   e

Figure 3.7   Examples of greaseless joints and taps.

(USA), a ridge is machined into the groove to increase the pressure on the
O-ring and reduce the chances of leakage. The elastomers most commonly
used for O-rings are hydrocarbon rubber (ethylene-propylene or butyl
rubber) and halocarbon rubber (e.g. Viton rubber) which swell when they
absorb solvents, but chemical incompatibility problems can often be
avoided by choosing the appropriate material. Hydrocarbon rubbers are
resistant to polar solvents and reducing agents whereas halocarbon
rubbers withstand non-polar solvents, oxidizing agents and Lewis acids.
Teflon O-rings and Teflon-coated O-rings are also available for these
types of joint, but they are less flexible and provide a less reliable seal.
Another method of reducing chemical contact with the O-rings is used in
Solv-seal joints produced by Fisher and Porter Co. (USA), although these
may no longer be available. In this variation, the O-rings are fitted onto a
Teflon cylinder which is pushed into the glass components as shown in
Figure 3.7c. The joint is then clamped.

   A variety of fittings which use threaded compression joints is also
available. We make extensive use of SVL fittings, one system of this type
which employs threaded glass tubing and plastic connectors in conjunction with Teflon-covered elastomer seals. Equal diameter tubes are joined
in a butt fashion (Figure 3.7d) while sliding joints are used for tubes with
different diameters (Figure 3.7e). For butt joints, the plastic couplings
comprise two linked threaded collars which can rotate independently,

while those for sliding joints are single piece mouldings. Sealing rings for the sliding couplings are also much thicker than those for the butt joints. By using different sealing rings, each threaded tube (available with diameters of 15, 22, 30 and 42 mm) can accommodate a range of smaller tube sizes. If you use this system, glassware incorporating threaded tubes should be stored with screw caps fitted to protect them from the rigours of everyday handling. These joints tend to chip fairly readily, and if the sealing surface is affected they cannot be repaired.

3.3.6   Attaching flexible hoses to glassware

At this point it is perhaps worth remembering that the techniques described here require the repeated connection and removal of flexible hoses to and from glass tubing. You need to become accustomed to the amount of force that can be applied safely and be able to do this perhaps a hundred or so times a day without breaking anything. When pushing rubber or plastic hose onto glass tubing, never use a twisting motion. The rubber can be eased over the glass with a slight rocking movement without using excessive force. If necessary, a small amount of silicone grease can be applied to the glass. With a plastic hose, it is sometimes best to warm it gently to soften the plastic. When removing a hose, ease it off by pushing against its end with the thumb of one hand while holding it and easing it gently from side to side with your other hand. Do not simply pull and twist with one hand. If the hose is reluctant to come off, use a sharp knife to cut it rather than risk broken glassware and a sliced hand.

3.4   Syringe and cannula techniques

Liquid transfer under an inert atmosphere is straightforward using either a double-ended stainless steel needle (cannula) or a syringe. Which you use for any particular transfer will depend on the volume and reactivity of the liquid to be transferred. Details of the procedures involved are given in the following sections.

3.4.1   Using cannulae

Cannulation relies on a pressure difference to transfer a liquid using a double-ended needle (Figure 3.8) or two needles linked by flexible PTFE tubing (Figure 7.3) and is the method of choice when large volumes are involved. If the pressure is provided from the nitrogen line, then the receiver must be isolated from the line and fitted with a bleed to achieve a pressure differential, unless you are using a nitrogen supply of the type shown in Figure 3.3b and described in section 3.2.2. For a standard

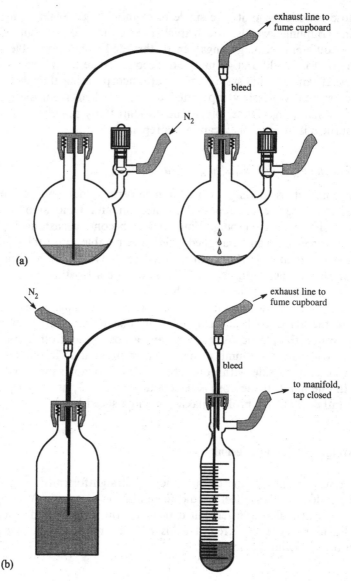

Figure 3.8 Cannula transfer of liquids.

manifold system, Figures 3.8a and 3.8b illustrate typical procedures described below.

- After evacuating and purging the receiving flask, remove the stopper under a steady nitrogen stream and insert a rubber septum cap of the correct size into the neck of the flask. To flush out the small volume of air in the septum cap, a short bleed needle is briefly inserted and then replaced by the transfer needle which is thereby flushed with nitrogen (if the cannula has only one sharpened end, this should go into the receiver). The bleed needle should be connected to the exhaust line, preferably through a bubbler containing less oil than the main bubbler on the line.

- The flask containing the liquid to be transferred is similarly fitted with a septum cap which is flushed with the aid of a bleed.
- The blunt end of the cannula is then inserted through a hole made with a sharp, non-coring needle (i.e. one which doesn't cut out a piece of rubber), and the bleed removed.
- Liquid is transferred by inserting a bleed into the receiver, closing the tap to this flask, and then pushing the blunt end of the cannula below the meniscus. If mercury bubbler pressure is being used, an excessive pressure build-up will force the septum cap out of the flask, but this can be prevented by watching the mercury bubbler and adjusting the nitrogen supply valve accordingly (and, if necessary, by wiring-on the septum cap). This problem will not occur when the tap to the oil bubbler is open.
- After the transfer, raise the ends of the cannula above the liquid levels, open the tap to the receiver, remove the bleed and then remove the cannula
- Clean the cannula immediately.

If you are using the more elaborate nitrogen supply shown in Figure 3.3b, then the higher pressure required for cannula transfer is established simply by closing the tap to the oil bubbler on the supply to the flask containing the liquid, while the receiver is connected to a tap providing a lower oil-bubbler pressure.

When transferring a liquid that is stored under an inert atmosphere in a bottle, pressure is provided by connecting the bottle to the line *via* a syringe needle through a septum cap. The nitrogen inlet from the line should first be inserted through the septum cap along with a bleed. Turn on the nitrogen at the line and purge the inlet before removing the bottle cap and quickly inserting the septum with nitrogen flowing. The bleed can then be removed before proceeding with the transfer as described above. Commercial reagents often come with their own seals, but remember that these have a limited lifetime, and Figure 3.8b shows how to transfer sensitive reagents into a calibrated tube for storage and subsequent cannulation of measured volumes.

3.4.2 Filtration with a filter stick

It is possible to modify the apparatus shown in Figure 3.8 to enable the separation of a solid product from the mother liquor by attaching a 'filter stick' to the cannula. The filter stick is simply a short piece of glass tubing with either a piece of hardened filter paper or sintered glass sealed over one end. If filter paper is to be used, it can be advantageous if some Teflon tape is first wrapped around the tube. Hold the paper flat against the end of the tube and, with finger and thumb of one hand, fold the overlapping paper up against the tube on two opposite sides and hold it firmly in place (Figures 3.9a and 3.9b). With the other hand, fold the remaining paper in a similar fashion (Figure 3.9c), press the filter paper against the Teflon and wrap a double-loop of tinned copper wire around it to hold it in place (Figure 3.9d). Grip the ends of the wire with pliers, twist while pulling gently to obtain a firm seal and then trim off excess wire and paper. In a variation on this design by M.L.H. Green the filter paper is supported on a short piece of flared capillary tubing which has been glued to a needle with epoxy resin (Figure 3.10) [1].

To use the filter stick, insert a septum cap of suitable size into the open end of the glass tube and pierce it with a sharp needle. Select rubber septa to fit the necks of the flasks to be used and push them onto the cannula with the stopper end towards the filter. Insert the blunt end of a cannula through the septum on the filter stick and carefully push it down close to the filter. Pump out a clean receiver flask (A, Figure 3.11), flush with nitrogen and insert the cannula (sharp end) and septum cap into the neck. The cannula and filter stick can now be purged for a few minutes. Push a bleed needle through the large septum cap to be inserted into the reaction flask (B). Under a nitrogen flow, remove the stopper from flask B and insert the filter stick and septum cap. After a few seconds remove the bleed. By inserting the bleed through the septum cap on the receiver (A) and closing the tap on this flask, liquid will be pushed through the

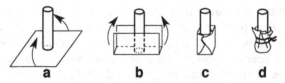

Figure 3.9 Construction of a 'filter stick' from glass tubing.

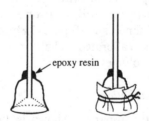

Figure 3.10 Attachment of filter paper to a purpose-built 'filter stick'.

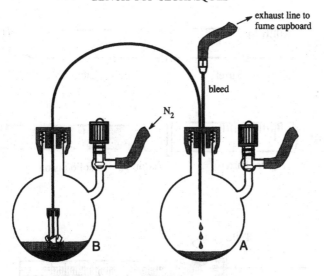

Figure 3.11 Filtration with a 'filter stick'.

filter when the filter stick is lowered below the meniscus in the flask B. It is best to allow solids to settle before filtering since fine particles can easily block these filters. Filtration can take some time, and it is always best to be patient. Note that if transfer is slow, then air may diffuse in through a bleed which is open to the atmosphere.

3.4.3 Syringes

A syringe fitted with a flexible needle is the most convenient method of transferring smaller, measured volumes of liquid (up to 50 cm^3), and you should be familiar with the various types and their limitations.

Types of syringe. Because of their solvent resistance, Luer syringes with glass barrels and pistons (Figure 3.12a) are used for most purposes, and several types are available. Barrels and pistons are often interchangeable but always check that these parts are marked as such before using a syringe of this type. Although adequate for most purposes, some leakage can occur between the barrel and the piston, and with air-sensitive reagents this will usually cause the piston to stick. When transferring a pyrophoric material, e.g. butyl lithium, this represents a significant hazard since there is no easy way of emptying the syringe of residual liquid. In such cases, it is preferable to use a glass gas-tight syringe having a metal or plastic piston with a removable Teflon tip (Figure 3.12b). Although much more expensive than the all-glass version, all of

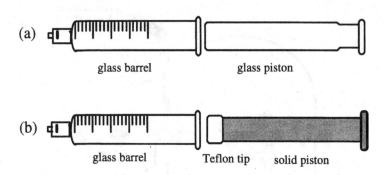

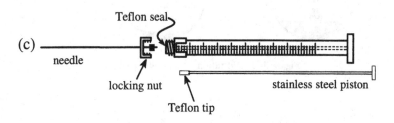

Figure 3.12 Syringes.

the components are replaceable and therefore have a longer life expectancy. A cheaper version containing a Teflon piston with an integral, Teflon-sheathed O-ring seal is also available.

Although disposable plastic syringes are considerably cheaper than the glass variety, there are problems with solvent compatibility. Where polypropylene is used for the barrel and the piston, the rubber sealing tip fitted to the end of the piston tends to swell in contact with organic solvents, causing it to stick. The use of slightly oversized rigid polyethylene pistons to create the seal against the polypropylene barrel overcomes this problem to some extent, although organic solvent compatibility should be checked before use. In general, it is best to reserve plastic syringes for use with aqueous solutions or for dispensing silicone or Apiezon greases.

Microsyringes offer the best method for accurate measurement of volumes less than 0.5 cm^3. The 100 µl size will probably be of most use, and the additional availability of a 10 µl version should cover most eventualities. Of the different types available, gas-tight versions with Teflon-tipped pistons and removable needles (Figure 3.12c) are preferable, since corrosion of the stainless steel pistons in liquid-tight designs can easily leave the piston locked solid in the barrel if the syringe is not cleaned thoroughly immediately after use. One irritating feature of microsyringes is that the needles supplied with them are usually quite short, although longer replacements can be obtained for the models with removable needles. Always bear in mind that these invaluable items are expensive and must be treated with appropriate care. The pistons are thin and easily bent; do not push quickly or with excessive force.

The standard-taper tips on Luer syringes can be either glass or metal and are machined to mate with matching sockets on the syringe needles. Glass tips snap off very easily and are best avoided unless a highly corrosive liquid is being handled. Luer-lock syringes are preferable, as these are fitted with locking rings so that needles can be secured by twisting into the screw thread in the locking ring. The best seals between syringe and needle are provided by Teflon Luer tips, although these are usually only fitted to gas-tight syringes. It is important to keep metal Luer tips and needle sockets clean to ensure a tight fit and prevent any ingress of air when the syringe is filled.

Syringe needles. The most commonly used syringe needles are constructed from stainless steel and have chromium-plated metal hubs (Figure 3.13), although Teflon needles with KEL-F hubs can be used where extra chemical resistance is required. The diameter of a needle is given as a gauge number, with smaller diameters having larger gauge numbers. The volume of liquid being transferred determines which gauge to use, and a rough guide is given in Table 3.2. Longer needles are preferred for liquid transfer to provide the necessary flexibility, while short ones are used as bleeds or for nitrogen inlets. It is important that the tips of syringe needles are kept sharp. Non-coring or deflecting

Figure 3.13 Syringe needle with non-coring tip.

Table 3.2 Needle gauges

Volume	Appropriate gauge	o.d. (mm)
5, 10 µl	26	0.40
25 µl–1 cm^3	22	0.64
1–5 cm^3	20	0.81
5–10 cm^3	18	1.02
10–20 cm^3	16	1.29
Large-volume cannulation	12	2.05

needles have 12° bevelled tips and cause less damage when pushed through septa.

Using syringes. You should aim to become adept at manipulating the syringe with one hand, leaving the other free to open valves, guide the syringe needle, etc. To avoid mishaps with solvents or liquid reagents, practice the following technique at your desk with an empty syringe until it no longer feels awkward.

- With your palm pointing inwards towards the syringe, use your index and middle fingers to hold the piston while gripping the barrel between the two lower fingers and your thumb.
- With index and middle fingers either side of the end of the piston, fill the syringe by extending the fingers (average sized fingers should enable about 20 ml to be withdrawn). The piston can then be prevented from sliding by holding it with your middle finger pressing against the end of the barrel.
- Expel the liquid by pushing against the end of the piston with your index finger. Don't push too hard, especially with interchangeable all-glass syringes, since liquid can be forced up past the piston to leak onto your hand.

Figure 3.14 shows the use of a syringe to transfer a measured amount of liquid from a storage bottle. Clean the syringe and needle thoroughly and dry in an oven at $>100°C$ before assembling the piston and barrel and securing the needle. While the syringe cools, fit the container with a septum cap and a nitrogen inlet, as described above for cannulation, and then follow the procedure below.

- First fill the syringe with nitrogen by inserting it through an existing hole in the septum cap, withdraw it and expel the nitrogen (Figure 3.14a). Repeat this flushing process two or three times.
- Push the tip of the needle below the surface of the liquid and draw a slightly larger volume than required into the syringe (Figure 3.14b).
- Raise the tip of the needle above the surface of the liquid and invert the syringe, pushing the piston in slowly to expel any gas until liquid squirts from the needle (Figure 3.14c).
- The volume in the syringe can then be adjusted to that required by either pushing in the piston or by submerging the needle tip and pulling out the piston to the appropriate mark on the barrel. It is important that the piston is now prevented from moving as described earlier.
- Grip the needle between thumb and forefinger of your other hand (about 5 cm from the tip), pull it out of the storage vessel and push it through the septum cap in the receiver (Figure 3.14d).

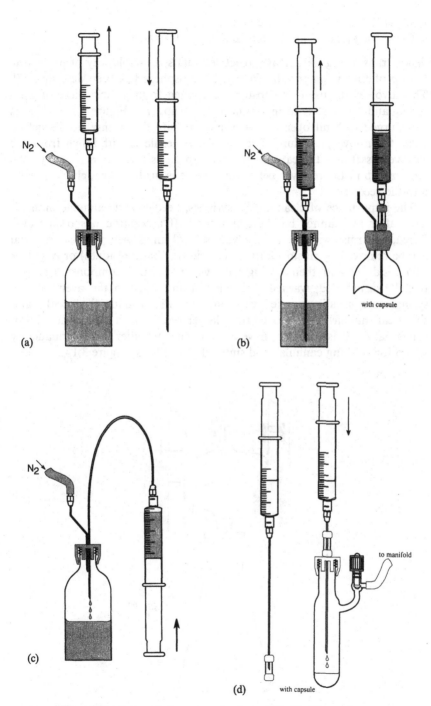

(a)

(b)

with capsule

(c)

(d)

with capsule

N₂

to manifold

Figure 3.14 Removal of a measured volume of liquid with a syringe.

- Expel the liquid into the receiver.
- Clean the syringe immediately and dry in the oven.

When transferring particularly reactive liquids, a simple way of protecting the tip of the syringe needle from the atmosphere has been described [7]. This involves the use of a capsule constructed from a short piece of glass tubing and two tightly fitting septa (details shown in Figure 3.14) which is first purged with nitrogen and then pushed onto the syringe needle before using the above procedure. The tip of the needle is withdrawn from the storage vessel into the capsule and the capsule is then pressed against the septum cap in the receiver before the needle is pushed through both septa into the receiver.

The importance of cleaning the syringes, needles and cannulae immediately after use cannot be overemphasized. If a solution has been transferred, first rinse with the neat solvent several times. Any residual material can be removed by rinsing with either acid (for basic residues) or base (for acid residues) and then rinsing with water. Syringe components should then be dismantled, washed with ethanol and dried in the oven. Microsyringes are usually supplied with fine wire for cleaning the needle, and solids can similarly be removed from larger needles with wire of the appropriate gauge. It is useful to have a large Buchner filter flask permanently set up for washing cannulae and sintered-glass filters (Figure 3.15).

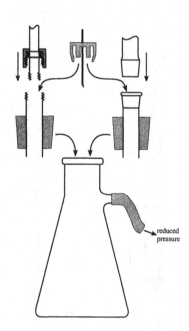

reduced
pressure

Figure 3.15   Buchner flask and adapters for cleaning glassware and cannula.

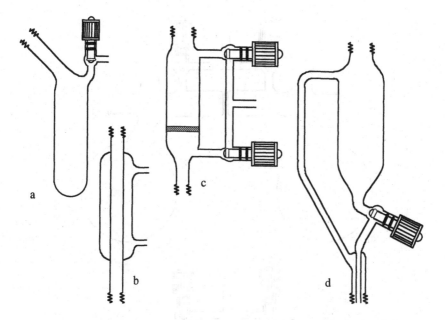

Figure 3.16 Assorted items of glassware for inert-atmosphere work.

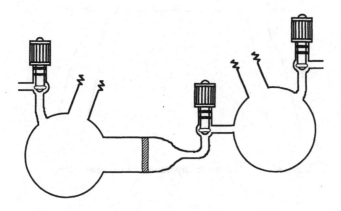

Figure 3.17 Reaction flask with an integral sintered-glass filter.

3.5 Apparatus

A basic set of Schlenk glassware will enable you to carry out the manipulations that will constitute a large part of your everyday routine, i.e. transfer of liquids and solids to a reaction flask, filtration of the reaction product, evaporation of volatiles and crystallization of the product. The

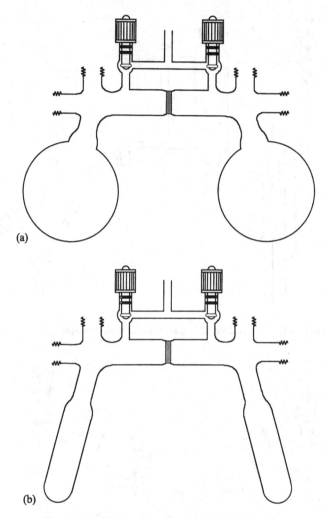

Figure 3.18   H-tubes incorporating sintered-glass filters.

most frequently used items of glassware are, therefore, flasks, solids addition tubes, condensers, filter tubes, solvent traps and the syringes and cannulae described in section 3.4. A selection of flasks has already been shown in Figure 3.6. Examples of apparatus with screw joints and grease-less taps are shown in Figures 3.16 to 3.18, but the same items can be constructed with ground-glass joints and taps.

Solids addition tubes (Figure 3.16a) are bent so that an upwards rotation about the joint causes the contents to drop into the reaction flask (usually with the assistance of some gentle tapping).

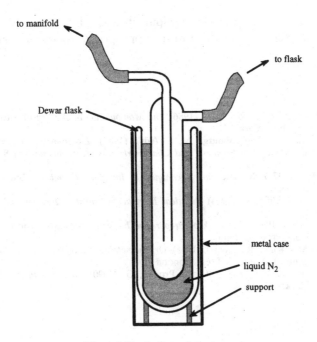

to manifold

to flask

Dewar flask

metal case

liquid N_2

support

Figure 3.19 In-line solvent trap.

A basic water-cooled condenser and a sintered-glass filter tube are shown in Figures 3.16b and 3.16c, respectively. The filter tube is designed so that the assembled apparatus, complete with the reaction flask and a receiver flask, can be used *via* a single tap on the manifold, which simplifies manipulations and avoids tangled hoses.

Whilst cannulae or PTFE tubing provide the most convenient means of adding a liquid to a flask, it is also useful to have one or two pressure-compensating dropping funnels (Figure 3.16d) for medium- to large-scale reactions.

Dual flasks incorporating a sintered-glass filter (Figure 3.17) eliminate a couple of joints and taps, and we find them convenient for medium-scale preparative reactions which produce solids. A related design is the H-tube described by Wayda [2] and versions which require only one hose connection to the manifold are shown in Figures 3.18a and 3.18b. The use of this type of apparatus with a single hose connection is described in section 8.2.1.

When removing volatiles from a reaction flask under reduced pressure, a solvent trap of the type shown in Figure 3.19 should be used. Note that it is connected to the flask and manifold as indicated to prevent blockage of the cooled inner tube.

A range of other items of equipment and glassware will also be required, and these are described in the appropriate section of chapter 7.

References

1. Shriver, D.F. and Drezdzon, M.A. (1986) *The Manipulation of Air-Sensitive Compounds*, 2nd edn, Wiley, New York.
2. Wayda, A.L. and Darensbourg, M.Y. (eds) (1987) *Experimental Organometallic Chemistry: a Practicum in Synthesis and Characterization*, ACS Symposium Series, Vol. 357.
3. Angelici, R.J. (1977) *Synthesis and Techniques in Inorganic Chemistry*, 2nd edn, W.B. Saunders Co., London.
4. Pass, G. and Sutcliffe, H. (1974) *Practical Inorganic Chemistry*, 2nd edn, Chapman & Hall, London.
5. Brauer, G. (1963, 1965) *Handbook of Preparative Inorganic Chemistry*, 2nd edn, Vols. 1 and 2, Academic Press, London.
6. Eisch, J.J. and King, R.B. (eds) (1965) *Organometallic Syntheses*, Vol. 1 (Transition Metal Compounds), Academic Press, London.
7. Casey, M., Leonard, J., Lygo, B. and Procter, G. (1990) *Advanced Practical Organic Chemistry*, Blackie, Glasgow.

4 Glove Boxes

4.1 Introduction

There are times when it is more convenient to handle air-sensitive compounds without their normal glass containment, for example when weighing out a pre-determined amount of solid or making a mull for IR spectroscopy. To do this you need a controlled, enclosed environment, usually in the form of a glove box, which is simply a large-volume gas-tight container from which oxygen and/or moisture are excluded (in reality they are kept at very low levels). If your work involves the use of a glove box on a routine basis, you should understand how it works and be able to operate it correctly. Such an expensive facility is likely to be communal, and an appreciation of the things that can go wrong will minimize down-time, not to mention friction within the group! The following sections are intended to cover these points in sufficient detail to allow individuals to operate a dry box safely, carry out troubleshooting and assess different designs if the purchase (or construction) of a new box is being considered [1].

4.2 General features

It is worth considering for a moment the stringent requirements which need to be met if a dry box (oxygen and moisture exclusion) is to operate successfully. Not only must the atmosphere inside the box be purified to bring the oxygen and moisture levels down as low as possible, but measures to minimize and counter contamination from outside the box must also be incorporated into the design. Given that a dry box will be used regularly for many years by a group of people that changes every year or so, a sturdy construction is also necessary.

These problems are tackled in slightly different ways by different suppliers and a range of designs is available commercially [2], including specialist versions for the industrial market which I won't describe here. All boxes intended for the research laboratory have the same basic features, while the (even more) expensive models include optional facilities such as automated gas handling systems.

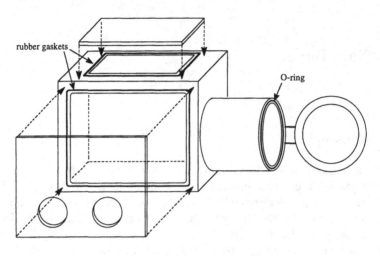

rubber gaskets

O-ring

Figure 4.1   Basic dry-box construction.

4.2.1 The box

The carcass of the box can be constructed from steel, stainless steel, aluminium or an acrylic polymer (polymethylmethacrylate). A typical metal carcass is shown in Figure 4.1. Steel boxes are epoxy-coated, whereas this is optional for the other metals. In modular designs, the end panels are removable so that boxes can be joined together to provide a larger working volume.

The basic glove box must be fitted with the following:

- an entry port for introducing items into the box without causing a major deterioration of the atmosphere;
- glove ports;
- windows to provide light and a clear view of the contents;
- breakthroughs for connections to external services.

The entry port. To minimize contamination of the nitrogen atmosphere within the box (note that argon or even helium are also used), items to be taken in are first placed in an ante-chamber (posting or entry port) which acts as an air-lock from which the air is removed and replaced with nitrogen from the normal supply by one of two methods.

(a) Leak-tight metal ports can be evacuated and then filled with nitrogen. As with Schlenk techniques, low levels of impurities can be achieved after several pump/fill cycles, at which point the inside door of the port can be opened. O-ring seals are used between the body of the port and the doors, which are usually attached by floating hinges and secured by evenly distributed perimeter bolts or a locking bar tightened with a central screw. To minimize evacuation times, a pump with as high a pumping speed as possible should be used. Entry times can be quite long if smaller pumps are used on large ports, preventing efficient use of the box. Smaller mini-ports are also available, and provide an efficient way to transfer smaller items into and out of the box.

(b) Alternatively, if the port cannot be evacuated, a continuous purge of nitrogen for an appropriate time can be used to reduce oxygen and moisture levels prior to entry into the box. High flow rates should be used to aid the interdiffusion of incoming gas with that already in the port, and the gas inlet and outlet should be sited so as to avoid a short-circuit.

Gloves. A compromise has to be made when fitting gloves to the box. Not everyone's hand size is the same, so the standard size gloves will be tight for some and loose-fitting for others. To add to the awkwardness, gloves diminish the sense of touch, although this varies with the type and thickness of rubber used. Butyl rubber is usually used because its permeability to oxygen and water is lower than that of neoprene. Natural rubber, although it has a better 'feel', is much more permeable than these synthetic materials. Two-piece plastic glove ports are sealed against the window with rubber gaskets and are either threaded (Figure 4.2a) or bolted (Figure 4.2b) together. The cuffs of the gloves fit over grooves in the outer ring of the port and are secured by O-rings which are clamped into the grooves by steel bands or plastic insulation tape. Bungs for the glove ports are often provided. These two-piece items usually pull into the port from inside the box and when tightened, an O-ring is squeezed against the inside wall of the port to create the seal. In some cases, an extra purging ring is inserted between the box and the external part of the glove port to enable connections to vacuum and nitrogen supplies (Figure 4.2c). With the bungs in place, the gloves can then be evacuated and flushed with nitrogen. This is particularly useful in the event of a punctured glove. Diffusion of water and oxygen through the gloves is one of the major sources of contamination, and sweat tends to accumulate in the gloves when you've been working in the box for any length of time. The discomfort can be alleviated by wearing thin cotton gloves or by applying talcum powder to your hands. Putting your hands into warm, damp rubber gloves after someone else has been working in the box is not the most pleasant of experiences!

Windows. Poly(methylmethacrylate) sheet of about 1 cm thickness is the usual window material, and rubber gaskets are usually clamped between

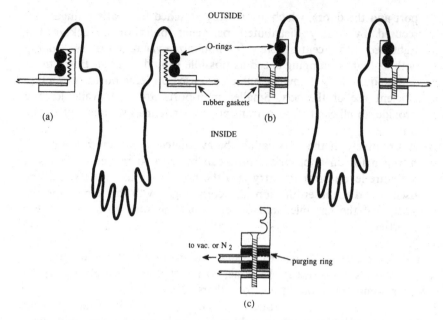

Figure 4.2   Glove port fittings: (a) screw threaded, (b) bolted, (c) with purging ring.

the metal walls and the windows to provide a gas-tight seal. The front window has holes to accommodate the gloves and a window in the top of the box provides light. Oxygen and moisture diffusion through acrylic windows is another source of contamination.

Breakthroughs. The gas inlet and outlet, connections to pressure gauges and gas bubblers and power supplies for electrical items in the box all need to come through the walls. These breakthroughs must be leak-tight and can be welded (or cemented in the case of acrylic boxes) or screwed into the metal wall. In some cases, two-piece fittings which screw together are used with rubber or elastomer seals each side of the wall.

4.2.2   Maintaining an inert atmosphere

Even with precautions to keep contamination to a minimum, the atmosphere inside the box needs to be either renewed or repurified. The simplest solution is to continually purge the box with nitrogen. This is discussed later, but a constant supply of gas is required and most commercial systems instead employ some means of recirculation.

Recirculation. Gas purification has already been mentioned in chapter 3, and water removal is most conveniently achieved by a column of 4A molecular sieves. Of the various oxygen scavengers which could possibly be used, commercial recirculation units generally use a supported-copper scavenger which is sold in pellet form as BTS catalyst. With a home-built system you can use the material of your choice, and we have used columns of supported MnO for some time on one of our dry boxes.

Dry box manufacturers use different ways of getting the gas through the absorption columns. With the columns outside the box, a unit employing circulating pumps is often supplied as a separate unit (Figure 4.3). The pumps and the joints in the pipework connecting component parts and leading to and from the box are a potential source of leaks in such a system, but these problems are greatly reduced in a simpler design where the gas is blown through an external purification column by a fan mounted inside the box (Figure 4.4). The columns in both of the above systems can be regenerated *in situ*.

By going one step further and putting the columns inside the box, the need for an external unit is eliminated, and this has been achieved in a fairly recent design from Belle Technology [3] by mounting the purifica-

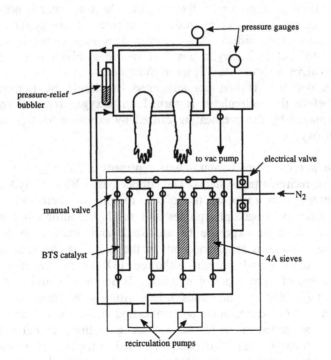

Figure 4.3 Schematic drawing of a dry box fitted with an external gas purification/recirculator system.

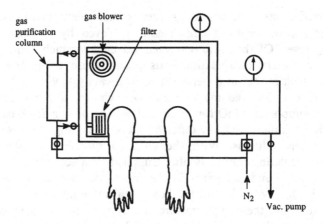

Figure 4.4 Schematic drawing of a dry box fitted with an internal blower and an external purification column.

tion columns directly onto a centrifugal fan (Figure 4.5). Although the columns must be sealed with special end-caps and brought out of the box for regeneration (a custom-built thermostatted heating band is provided), the overall simplicity and comparatively low price of this system should prove attractive for academic laboratories. This recirculator is supplied with the manufacturer's acrylic box, but is also available separately for those who might wish to install it in an older box.

When a dry box is first commissioned, it should be purged with nitrogen before the recirculator is turned on. Oxygen removal columns can be damaged by the overheating caused by exposure to high concentrations of oxygen.

Continuous purge. Some contaminants represent a threat to the oxygen scavenger in recirculating systems. The copper based BTS catalyst cannot be regenerated after exposure to volatile organic or metalloid halides, halogens, mercury, organophosphines and sulphides, although regeneration is possible after exposure to ammonia and amines. In research groups where reactions are carried out in the box, this can limit the use of various solvents and ligands, although 13X sieves are effective in removing some of these types of materials. It has been pointed out that, for moderately sensitive compounds, a box with a slow, continuous purge of nitrogen and no recirculation or purification performs well and allows reactions to be carried out inside the box using routine manipulations [4]. Note that if noxious compounds are to be used in the box, it is especially important that vented gases are led to a fume cupboard and that the entry port is in a well-ventilated area.

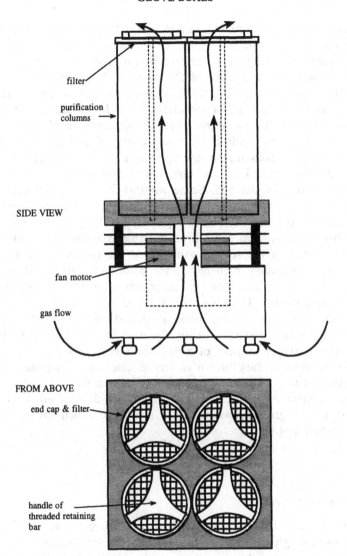

Figure 4.5 Stand-alone combined fan and gas purification unit for use inside a dry box.

Monitoring the purity of the atmosphere. Although commercial oxygen and moisture meters are available, they tend to be expensive and it is therefore not universal practice in academic laboratories to use them for automatic monitoring of the dry box atmosphere. More often, it is the untoward behaviour of air-sensitive samples that alerts users to a deterioration in the atmosphere. Qualitative tests are often used to check the box before risking exposure of samples, one of which uses the filament lifetime of a standard tungsten filament light bulb with a hole cut in the glass envelope. For the bulb to remain lit for days to weeks, the total oxygen plus water level must be of the order of 5 ppm or lower. Observations of oxygen or moisture-sensitive reagents provide an immediate visual indication of the state of the atmosphere. For example, $TiCl_4$ will fume if the moisture level is above about 10 ppm and metal alkyls (most commonly $AlEt_3$ and hydrocarbon solutions of $ZnEt_2$) fume when exposed to oxygen levels in the low ppm range. The drawback of these tests is that they contaminate the atmosphere and leave deposits on the walls and windows of the box. Some workers use a solution generated by reducing Cp_2TiCl_2 with zinc metal powder in dried, deoxygenated tetrahydrofuran, benzene or toluene [5]. Drops of this green solution are placed onto an aluminium sheet or disposable weighing pan and if the colour is retained as the solvent evaporates, the oxygen level is below 5 ppm. With increasingly high levels, the colour changes to olive-green or yellow and eventually orange, in which case the atmosphere is seriously contaminated. However, this method again introduces solvent vapours into the box. It has been suggested that a small tube of manganese(II) oxide on silica gel could be used to provide a quantitative measure of the oxygen content of a dry box atmosphere [1]. The colour change from green to dark brown is very distinct and, since the loading of manganese onto the silica gel is known, the amount of oxygen in a measured volume of gas could in theory be determined from the movement of the green/brown interface given the stoichiometry of the reaction (equation 4.1).

$$2MnO + \tfrac{1}{2}O_2 \longrightarrow Mn_2O_3 \qquad\qquad (4.1)$$

4.2.3 Pressure control

The pressure inside the box should be slightly above 1 atm, otherwise it becomes increasingly difficult to use the gloves. In addition, pressures much above or below this can damage the box so it is important to have a control system that can handle large fluctuations. During use, the internal pressure increases as you push your arms further into the box. A maximum pressure is set either by the level of oil in a bubbler attached directly to the box or by a pressure switch that controls an electrically operated exhaust valve (a bubbler set to a higher pressure is also fitted in this instance in case the electrical system fails). Similarly, when the gloves are withdrawn, the pressure decreases. When the minimum pressure is reached (this is set by adjusting a pressure switch), nitrogen is admitted to the recirculation system or directly to the box if a continuous purge is being used. In commercial dry boxes, the maximum and minimum pressure switches are often incorporated into a single pressure controller. The inlet valve is usually controlled so that the pressure increases to a set level between the minimum and maximum before it closes. Some models incorporate a safety cut-out switch that turns off all power to the box if the pressure reaches a pre-set level above the maximum determined by the bubbler. This alarming event may indicate a malfunction, but it is more likely that both arms have been pushed into the box too quickly for the gas to escape through the bubbler. Once the pressure has dropped, the switch can be reset to restore power.

4.3 Maintenance and use

Glove-box technology was originally developed by the nuclear industries to provide the high levels of containment that are necessary. In chemical research laboratories the nature and extent of their use varies widely. It has been the practice in several large groups in the USA for researchers to use dry boxes for almost all operations. This means that the boxes are in constant use, and one hears rumours of the introduction of shift work to enable round-the-clock operation! In continental Europe, the tendency has been to avoid dry boxes and use Schlenk and vacuum-line techniques. We have adopted the most common practice of using dry boxes for weighing, transfer and storage and the more intricate operations, while most reactions are carried out using Schlenk techniques.

The trepidation with which most students first approach a dry box can be avoided if some knowledge of the principles of design and operation is acquired beforehand. This chapter hopefully will instil a modicum of confidence, but it is worth reading the instruction booklet for the particular model (if it can be found). Most calamitous situations can be prevented by thinking ahead and following a basic set of rules and procedures.

4.3.1 Preparing to use the dry box

When you need to use the dry box, think through all the manipulations you will carry out and try and foresee all eventualities. Make sure that you take into the box everything you are likely to need; it is very frustrating to go through the entry routine only to find that you've left a crucial item outside and have to repeat the whole procedure (a mini-port can be a blessing in such circumstances). If there is no balance in the box and you want to transfer a known amount of a substance in the box, you need to weigh the empty flask and its stopper under nitrogen first. Label the component parts or ensure that there is no chance of mixing up stoppers in the box!

If the box has an evacuable port, check the joints of all the flasks to be taken in. Any flask with ground-glass joints **must** be evacuated at the bench; otherwise the stopper will fly out and break when the port is evacuated and may damage other glassware in the process. This type of flask must never be used to take liquids into the box through an evacuable port, and solids must be pumped dry to prevent the stopper from being blown out by the vapour pressure of residual solvent. (It is possible to secure ground-glass stoppers with electrical tape to enable flasks at atmospheric pressure to be taken through an evacuable port.)

These precautions are unnecessary for screw-capped flasks or apparatus with clamped O-ring joints, and they can be used to take liquids into the box *via* an evacuable entry port. Similarly, if the box has a purged entry port the above restrictions do not apply.

As far as is possible, all items should be clean and oven-dried. Closed containers should be purged or evacuated to eliminate contamination of the atmosphere inside the box. Porous, absorbent materials such as wood, paper and corroded metal contain water and should be avoided if possible. Where it is felt necessary to take paper towels into the box, they should be dried beforehand in a vacuum desiccator over a drying agent or in a vacuum oven and then pumped in the entry port overnight.

4.3.2 Taking items in

Having collected together everything you need, it helps to have a container for all the items. We find that the larger plastic margarine or ice-cream tubs (the square or rectangular ones) are a convenient size and, in addition to making it easier to load and unload the port, they can protect the floor of the box from spillages while you work. Before opening the outer door of the entry port, check to make sure the inner door is closed properly. You can then unclamp the outer door and place the items in the port, preferably near the inside door so they are easier to reach from inside the box.

Evacuable ports. Close the outer door and tighten it just enough to compress the O-ring seal. If perimeter bolts are used, tighten the nuts evenly; otherwise the door may become distorted. The port can then be evacuated.

• If it is not already on, turn on the vacuum pump and then open the vacuum valve on the port. Check the pressure gauge and if the port is not being evacuated, adjust the clamps on the outer door until it is. It is also wise to check the pressure inside the box. If it decreases when the vacuum pump is turned on, the clamps on the inside door will need to be tightened. Do not overtighten a door while the port is evacuated – this will cause damage and make it difficult to open the door at atmospheric pressure.

- When the pointer of the pressure gauge has stopped moving, close the vacuum valve.
- Open the nitrogen inlet valve and again watch the pressure gauge on the port. Before it reaches atmospheric pressure, close the valve to prevent the pressure breaking the door seals.

Repeat this procedure another two or three times (this is done automatically on some models). During the last purging cycle, the port can be pumped down for longer (several minutes) before filling with nitrogen. If a connection is provided between the port and the box it can now be opened to equalize the pressures; otherwise nitrogen is admitted to the port until the door opens easily.

Purged ports. Usually fitted to acrylic boxes, these have doors similar to glove-port bungs, sealing against the inside of the port when a central nut is tightened. Once the items are inside the port and the door is secured, the nitrogen purge can be turned on. Flow rates of about 30 l min^{-1} and purging times of 3–5 min are recommended for the acrylic box in our laboratory. After this time, the purging valve is turned off and the inside door can be opened.

4.3.3 Working inside the box

When you are ready to open the inner door, put your hands into the gloves. (This simple instruction hides the effort that is often required to follow it, especially if the gloves are reaching out to greet you because the box pressure is slightly above atmospheric and no bungs are in place.) Start with one hand, pushing the fingers inwards and gradually working your whole hand into the glove. It is easier if you push the gloves into the box to expel some nitrogen and reduce the pressure. Whenever possible, try to keep the pressure constant by pulling one arm out when reaching into the box with the other. This will minimize large pressure variations and prevent repeated cycles of gas expulsion followed by re-pressurization.

Once the items have been removed from the entry port, the inner door should be closed just in case anyone inadvertently opens the outer door or it isn't properly sealed. It is unlikely that the gloves will be a perfect fit. Over-sized finger-ends tend to get in the way and you must be careful not to cut or pierce them, especially when opening ampoules or using syringes. When transferring material in the box, always work over a removable tray or plastic container and bring out any spilt material, dirty pipettes or broken glass. Any solid that falls onto the dry-box floor can be wiped up with Parafilm. Static charge develops readily in the dry atmosphere of the box, but a piezoelectric antistatic pistol can help prevent powders or small crystals from flying everywhere except where you want them. The most convenient method of weighing air-sensitive materials is to have a balance in the dry box. If the balance is too large to be taken in through the entry port, a panel or window must be removed. This is obviously best done when the box is being set up and contains no sensitive compounds. Samples are often brought into the box to be transferred to glass ampoules for sealing under nitrogen (chapter 8). If it is necessary to know the amount of sample in each ampoule and there is no balance in the box, the ampoules and taps must all be pre-weighed and labelled to avoid any mix-up in the box. They are then taken out of the box, sealed off and the component parts re-weighed.

For my money, one of the most challenging feats in a dry box is the selection and mounting of single crystals for X-ray structure determinations. Transferring a crystal on the end of a glass fibre into a Lindemann capillary while your hands are in oversize rubber gloves is a supreme test of patience and, although there are now more convenient ways of mounting crystals, perhaps everyone should go through this character-building exercise at least once in their career!

4.3.4 Taking items out

When you are ready to exit the box, tidy up. Only store samples in the box that are likely to be needed in the foreseeable future; otherwise transfer them to glass ampoules and seal them for storage outside the box. After placing all items in the port, ensure that the inner door is properly tightened and sealed before removing your hands from the gloves. If the glove port bungs are to be inserted, pull them into place one at a time (you may need to slacken the central nut slightly), rotating each one to fit behind the retaining tabs, and then tighten the central nut to seal it firmly in place.

Should you ever need to remove an item without first having entered the box, remember that the entry port must be pumped and purged as described above before the inner door is opened.

4.3.5 Useful items

To make full use of a dry box facility, it is helpful to keep a range of equipment inside the box. Shelves and small plastic stackable bins are ideal for storage. Retort stands and clamps or a small lattice are invaluable for supporting glassware. A balance has already been mentioned and if other electrical items are to be used in the box then a small electrical distribution adapter will be necessary (note that metal parts of the box should be properly earthed). If reactions are to be carried out in the box, then a magnetic stirrer and followers will be required in addition to a collection of flasks and filters. A vacuum pump and solvent traps can be connected through a breakthrough to a tap inside the box to enable filtrations and solvent removal under reduced pressure. Small freezers are also available for installation inside the box (although they are not always reliable) and are useful for recrystallization. For transferring liquids, it is convenient to keep a selection of syringes in the box, in addition to disposable dropping pipettes with rubber teats. NMR solvents and dry mineral oil will be needed on a routine basis for the preparation of spectroscopic samples. Aluminium foil boats are useful for weighing solid samples and transferring them into containers is easier with the aid of a small glass funnel. It is also wise to have a collection of spatulas and tweezers to hand. In larger boxes where items at the back, in corners or on shelves are difficult to reach, a pair of tongs may aid their retrieval.

4.3.6 General maintenance

It is advisable to keep a log book on or near the dry box. In addition to identifying the last person to use the box (assuming everyone is honest) this provides a record of leaks, regeneration times, glove changes, etc. For average use, the molecular sieves should be regenerated about once every month or so whereas, with careful use, the deoxygenation columns could last for up to a year. It is important that the box be kept as clean as possible and the O-ring seals on the entry port doors should be checked regularly and wiped clean. The acrylic window, which tends to become greasy due to the foreheads and noses of users who forget about its existence, can be wiped clean with a little detergent but **do not use organic solvents**. Recirculation pumps become noisy with age as bearings become worn and these can be replaced along with the diaphragms.

References

1. For a more detailed discussion of dry box design, see Shriver, D.F. and Drezdzon, M.A. (1986) *The Manipulation of Air-Sensitive Compounds*, 2nd edn, Wiley, New York.
2. Major glove box manufacturers include:
   Vacuum Atmospheres Co. at: 4652 West Rosencrans Avenue, Hawthorne, CA 90250, USA and at: Justinianstraße 22, Lairco-Haus am Holzhausenpark, D-600 Frankfurt/Main 1, Germany.
   M. Braun at: Gutenbergstraße 3, D-8046 Garching b. München, Germany and at: 2 Centennial Drive, Suite 4F, Peabody, MA 01960, USA.

Innovative Technology Inc., 2 New Pasture Road, Newburyport, MA 01950, USA.
Saffron Scientific Equipment Ltd., Firtrees, Sires Hill, Didcot, Oxfordshire, OX11 9BG, UK.
3. Belle Technology, Unit 1, Walnut Orchard, Portesham, Dorset, DT3 4LH, UK. (This company manufactures acrylic dry boxes, an in-box recirculator and associated column regeneration equipment.)
4. Marder, T.B. (1987) In *Experimental Organometallic Chemistry: a Practicum in Synthesis and Characterization*, eds A.L. Wayda and M.Y. Darensbourg, ACS Symposium Series, Vol. 357, p. 153.
5. Sekutowski, D.G. and Stucky, G.D. (1976) *J. Chem. Educ.*, **53**, 110.

5   Operations on the High-Vacuum Line

5.1   Introduction

The chemical vacuum line was developed in Europe during the early years of this century and Alfred Stock's original design for handling boranes has since been modified and adapted by many research groups. The high-vacuum line provides the ideal enclosed air-free environment for reactive chemicals. It is most often used for reactions involving a gas or a reagent with a high vapour pressure, whereas Schlenk techniques are used for solution chemistry on the bench top. In this chapter, the basic principles of the design and operation of a high-vacuum line are followed by more detailed descriptions of the manipulation of volatile, air-sensitive chemicals, and then a section on different types of high-vacuum pumps. If your work is likely to involve extensive use of a high-vacuum line, I recommend that you also read at least one of the publications containing excellent detailed descriptions of their design and use [1,2].

5.2   The high-vacuum line

The quantitative transfer of volatile compounds is readily achieved on a high-vacuum line, and the basic principle is illustrated schematically in Figure 5.1. The liquid in A can be transferred to B by distillation, but the diffusion is hindered if molecules (or atoms) of a non-condensable gas are present, so these must first be removed. Container A is therefore cooled to $-196°C$ in liquid nitrogen and extraneous gases are pumped out by evacuating the system to pressures of $10^{-3}–10^{-6}$ mmHg, lower than the $10^{-1}–10^{-3}$ mmHg used in the Schlenk manifolds discussed in chapter 3. Once the sealed system has been warmed to ambient temperature, only the liquid and its equilibrium vapour pressure are present (any non-condensable gases dissolved in the liquid must also be removed by freeze–pump–thaw cycles as described in section 5.3). The vapour can now be condensed into B by cooling this receiver in liquid nitrogen while gently warming A.

Depending on the nature of the compounds to be handled and the type of manipulations required, the design of a high-vacuum line can range from a simple, bench-mounted single manifold to a complex multipurpose system which is usually mounted on a lattice fixed to a rigid frame with a

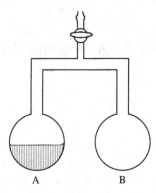

Figure 5.1   Basic vacuum-transfer arrangement.

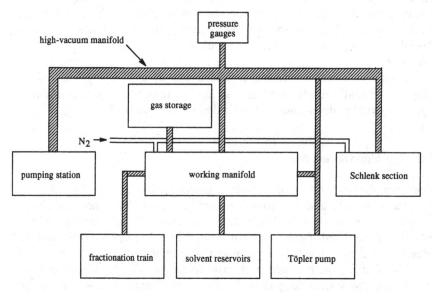

Figure 5.2   The elements of a multipurpose high-vacuum line.

low bench to provide sufficient space for the attachment of various
apparatus. Designs are often inherited and then modified to suit particular projects (or budgets) and as such can become quite personalized. A
general layout for a typical multipurpose high-vacuum line installation is
shown in Figure 5.2, and the following discussion of the component parts
gives some idea of the different approaches to vacuum line work that may
be encountered.

5.2.1   The pumping station

The low pressure necessary for the line is provided by a pumping station which requires a rotary backing pump (section 3.2.3) connected in series with either a diffusion pump or a turbomolecular pump (section 5.4). A schematic representation is shown in Figure 5.3 (note that the tubing connecting the high-vacuum pump to the manifold should be as wide as possible to maximize the pumping speed). The parts within the dotted outline are only necessary if a diffusion pump is installed; an additional trap to collect mercury or oil backstreaming from the diffusion pump, and a by-pass to protect the diffusion pump from higher pressures during operation. The exhaust from the rotary pump should be led to the fume cupboard.

Remember to **avoid** condensing liquid oxygen in traps cooled by liquid nitrogen. Not only does it create an **explosion hazard** when in contact with oxidizable organic substances, but its evaporation upon sudden warming will pressurize the vacuum line with **potentially disastrous consequences**. Check that there are no stopcocks open to the atmosphere, and **never** expose the cold traps to the atmosphere for any length of time.

With the traps immersed in liquid nitrogen, initial evacuation of the line is achieved using the rotary backing pump with the diffusion or turbo pump turned off. Once the pressure is below 1 mmHg, the diffusion/turbo pump can be turned on (together with the water supply if a water-cooled

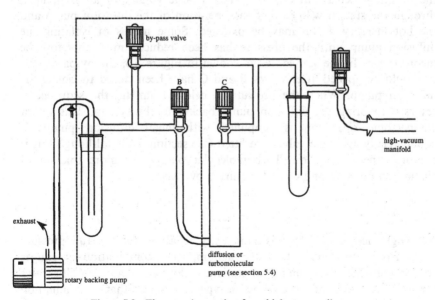

Figure 5.3   The pumping station for a high-vacuum line.

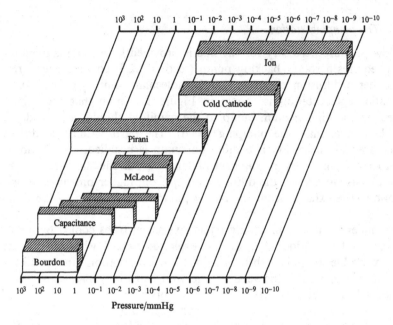

Figure 5.4 Pressure ranges for various types of pressure gauge.

diffusion pump is being used!). In a system containing a diffusion pump, the resulting increase in pressure when a large volume of air is pumped through the station will prevent the operation of the diffusion pump and the hot mercury or oil may be oxidized. Some means of isolating the diffusion pump until the pressure has been reduced must therefore be incorporated. In the simple design shown in Figure 5.3, the by-pass valve A should be opened after valves B and C have been closed to isolate the diffusion pump. When the pressure is below 1 mmHg, the sequence is reversed to bring the diffusion pump back into the system. This rather cumbersome operation is simplified with metal diffusion pumps if commercially available valves are fitted (see section 5.4.1), although this is a more expensive option. Turbomolecular pumps are more tolerant of changes in pressure and do not require a by-pass.

5.2.2 Pressure gauges

No single design of gauge is sufficiently reliable over the full pressure range from atmospheric to high vacuum, so a combination of gauges covering smaller, overlapping ranges is usually used. Figure 5.4 illustrates the working ranges of some common types and shows that a combination of Pirani and cold-cathode gauges will cover the pressures encountered in chemical vacuum lines. For quantitative manipulations of volatile substances using pressure–volume relationships (manometry), either mercury manometers, capacitance manometers or Bourdon gauges are needed to measure pressures between 1000 and 1 mmHg, while thermionic gauges are only necessary with ultrahigh-vacuum apparatus. The basic features of the various gauges are discussed below.

Bourdon gauges. These mechanical devices (Figure 5.5) consist of a thin-walled metal spiral enclosed in an envelope which provides the reference pressure, usually atmospheric. Internal pressure variations cause the end of the spiral to move, and these deflections can be measured and calibrated. The same principle applies in glass spiral gauges where a delicate coil of glass tubing is mounted inside the reference envelope which is connected to high vacuum to provide a convenient reference pressure.

Capacitance manometers. In this type of pressure transducer, small deflections of a metal diaphragm are measured by detecting changes in capacitance (Figure 5.6). These devices give a digital readout and are designed to cover different pressure ranges, as indicated in Figure 5.4. They have the advantage of being fairly robust and chemically inert and pressure readings are independent of the type of gas, although ambient temperature variations can affect long-term stability (some designs incorporate heating elements to overcome this problem).

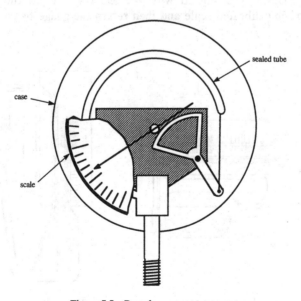

Figure 5.5 Bourdon pressure gauge.

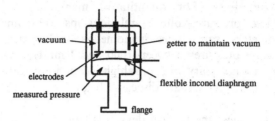

Figure 5.6   Schematic diagram of a capacitance manometer.

McLeod gauges.   These mercury-filled gauges operate by compressing a known volume of a low-pressure gas until the resulting pressure is sufficient to be measured by a column of mercury. The commercially available Vacustat is a miniature, tilting McLeod gauge (Figure 5.7), sold with a stand for bench-top use, or without a stand for permanent connection to a vacuum system. The gauge is connected to the system *via* a tube extending backwards from the diagram (A) which is fitted with either a nipple (for portable use) or a ground-glass joint (for direct connection to the line). To prevent broken columns of mercury becoming lodged in the sealed capillary (B), keep this type of gauge in the horizontal position except when actually measuring the pressure. To measure the pressure, tilt the gauge until the mercury enters capillary B and the mercury in the reference capillary (C) is aligned with the top of B. Read the pressure directly from the calibrated scale **and then return the gauge to the horizon-**

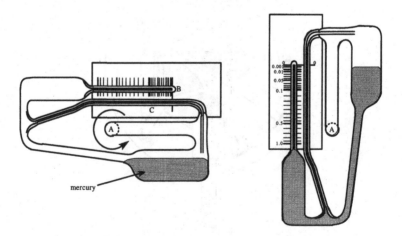

Figure 5.7   Tilting McLeod pressure gauge ('Vacustat').

tal position. The amount of mercury required by the gauge is often etched on the glass for refilling after cleaning.

The more traditional design (Figure 5.8) is permanently connected to the vacuum line, contains a much larger amount of mercury and is less easy to use. By admitting air to the reservoir (A), mercury is allowed to rise slowly into the gauge, and when it passes the point indicated by the arrow, gas is trapped in the measured volume which includes capillary B. If the level of the mercury is adjusted until the top of the mercury column in the reference capillary (C) is aligned with the end of the sealed capillary B the pressure of the compressed gas (in mmHg) is then the difference between the levels of mercury in C and B and the initial pressure of the gas can be read from the calibrated scale. A more detailed description of the principles involved is given below which also includes a modified measuring procedure that enables a linear scale to be used.

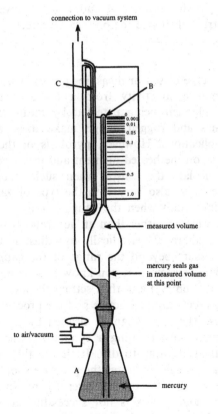

Figure 5.8 Traditional McLeod pressure gauge design.

Theory

If the initial pressure of the gas is P, the known volume of the bulb and capillary B above the point indicated in Figure 5.8 is V, the compressed gas occupies a volume v at a pressure p in capillary B (area A and length h (from the sealed end to the mercury meniscus)) then:

$$PV = pv$$
$$P = h^2 A / V$$

Since A and V are known, P can be calculated from h although a pre-calibrated scale is usually provided with commercial gauges.

Alternatively, the gas can be compressed into the same length of capillary h_0 every time, in which case:

$$PV = pv_0$$
$$P = h_0 A (\Delta h) / V$$

where Δh is the difference in meniscus heights between the mercury in C and that in B. In this case h_0, A and V are known and a linear scale can be constructed to give the pressure P directly from Δh.

Pirani gauges. This type of sensor measures pressures by determining the heat loss *via* thermal conductivity from a heated wire filament, the temperature of the filament being detected by means of a thermistor. Rapid response times and rugged design make these gauges ideal for general purpose applications. However, pyrolysis of thermally sensitive compounds can occur on the heated filament and it is a good idea to turn off the gauge and isolate the sensor when such materials are being handled. Pressure readings also vary with the type of gas in the system, although this isn't important when the gauge is only being used to give an indication of the vacuum rather than an accurate measure of pressure. The gauges must be calibrated periodically by adjusting the zero reading at a suitably low pressure beyond the range of the gauge (this assumes that you have a vacuum system that you know is working well and which will provide about 10^{-5} mmHg) and then setting the atmospheric pressure reading. These adjustments are interactive and the process may have to be repeated a few times. The sensor can be cleaned by gently filling with a non-chlorinated organic solvent using a syringe, leaving for at least 10 min, draining and then allowing to dry. It should then be evacuated to below 10^{-4} mmHg. After a few minutes the pressure should drop from an initially high reading (in the 10^{-2} range) to zero. If not, a zero adjustment must be carried out, and if this is not successful the sensing element should be replaced.

Cold-cathode and thermionic gauges.   These devices are used for pressures below 10^{-3} torr, and measure the ion current when residual gas molecules are ionized by bombardment with electrons. In the cold-cathode gauge, a very high voltage pulse initiates a low pressure discharge. The concentration of electrons is maintained by confining them to spiral paths in a magnetic field, thereby increasing the number of collisions with gas molecules and sustaining the discharge under an applied high voltage. A typical design which incorporates a graphite cathode is shown in Figure 5.9. As with the Pirani gauge, sensitivity is dependent on the nature of the gas in the system, and when significant pressures of polyatomic molecules are present it is best to turn off the gauge since decomposition products tend to accumulate in the sensor. Despite this, the robust construction and tolerance to deposits in the sensor make this an ideal gauge for high-vacuum measurements on a chemical line and when it is hard to start or readings become erratic the sensor can be dismantled and cleaned. The glass insulator and internal surfaces should be washed with acetone followed by alcohol, while sputtered deposits on the anode can be removed with an abrasive cloth. Graphite cathodes should not be cleaned, but can be replaced if badly worn away due to sputtering.

Hot-cathode or thermionic gauges usually have the Bayard–Alpert configuration in which electrons emanate from a heated filament and are accelerated towards a positively charged grid by a high voltage. In the space between the filament and the grid, gas molecules are ionized and the positively charged ions then pass through the grid to be collected at an electrode which is held at a negative potential with respect to the filament. As in the cold-cathode gauge, the ion current is related to the pressure and the response is dependent on the type of gas. Polyatomic

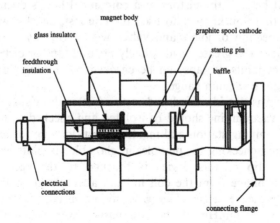

Figure 5.9   Cross-section of a cold-cathode pressure gauge.

molecules decompose readily at the heated filament so these gauges are preferred for clean, high- and very high-vacuum systems rather than those used for synthesis.

Pressure gauge controllers.   The signals from several transducers are often fed into a single electronic display/control unit which can be programmed to turn the various sensors on and off at appropriate pressures and thereby minimize the chances of damage. Sensor heads which incorporate their own processing electronics (active gauges) enable supply and output voltages for different types of gauge to be standardized, are far less susceptible to electronic noise and make remote operation with long cable runs possible.

5.2.3   The vacuum manifold

The high-vacuum in a multipurpose line is supplied to all the sections *via* a single manifold fitted with pressure gauges and constructed of fairly wide-bore tubing to maximize pumping speed. To impart some flexibility, connections to the pumping station and to the various other sections can be made with ground-glass (greased or waxed) or O-ring ball and socket joints. Every joint and valve is a potential source of a leak, so the design should be kept as simple as possible while catering for intended uses.

After turning on the line, gases and solvent vapours are removed from stopcock grease, O-rings and glass surfaces and it may take some time to pump down to $<10^{-3}$ mmHg. Leaks are a ubiquitous source of frustration with high-vacuum lines so be prepared for some detective work if, as often occurs, the pressure is still relatively high even after pumping overnight. It is important to be systematic and thorough in your search for leaks and it helps if the history of a communal line is documented in a log book so that you know who used the line last, and for what it was used. Close inspection of joints and valves will reveal any obvious leaks due to channels in the grease, inadequately greased taps or dirty/damaged O-rings. Suspect joints or taps can be checked by turning them slightly while observing the vacuum gauge.

If the problem is not identified and rectified after the visual inspection, sections of the vacuum line should be isolated and checked systematically. Scrutinize the pumping station and manifold first, looking for any volatile contaminants and checking for pinholes in glass-blown joints with a Tesla coil. When the tip of this device is touched to the glass, the high frequency high voltage from the coil induces a purple glow discharge in the line in the presence of air which will only be sustained if the pressure is greater than 10^{-3} mmHg. If the coil passes over a pinhole as it is moved over the glassware, a blue spark will appear at the site. Note that a Tesla coil should not be used near thin glass or O-ring seals (punctures may result) and it is not likely to be useful around greased joints. Metal objects should also be avoided since they simply serve to ground the Tesla coil.

5.2.4 The work station

This portion of the line provides the means to connect reaction flasks to the high-vacuum manifold, to solvent and gas reservoirs, to inert gas and reagent inlets, to low temperature traps and manometers and, if necessary, to a Töpler pump for the quantitative transfer and measurement of non-condensable gases. This is the heart of a chemical high-vacuum line and can be quite complex although not all facilities will necessarily be included in any particular installation. Figure 5.10 shows how the various components may be interconnected. The valves may be either high-vacuum greased stopcocks, greaseless taps or a combination of both, but bear in mind the relative strengths and weaknesses of the different types as highlighted below.

- Compounds are absorbed by grease and their rate of desorption varies. In some cases, reactions occur which cause taps and joints to stick.

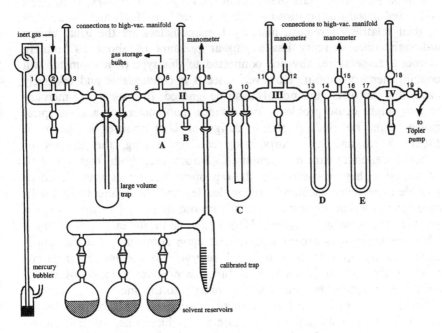

Figure 5.10 Typical high-vacuum work station.

- Greased taps tend to 'streak' with use and the channels in the grease shorten the leakage paths for air into the system, especially with oblique taps.
- Hollow ground-glass keys which are pulled into the tap under vacuum need to be greased (thereby exposing the inside of the line to air) more frequently to prevent them from sticking.
- The operation of greaseless Teflon screw-valves is easily impaired by small amounts of solid on the valve seat, and they must be used carefully and kept clean to avoid leaks.

In Figure 5.10, reaction flasks are attached to section **II** at sites such as **A** or **B**, which can accommodate different joint types. With section **II** isolated (taps 5, 6, 8 and 9 closed), measured volumes of solvents can be added from the solvent manifold *via* the calibrated trap. Alternatively, solvent reservoirs may be attached directly to section **II**. With taps 1, 2 and 6–9 closed, solvents can be pumped from a reaction flask and collected in the large cold trap connected between sections I and **II** by opening tap 3 (volatile liquids should not be allowed to enter the high-vacuum manifold). Nitrogen can be admitted into sections I and **II** and hence into the reaction flask by closing tap 3 and opening taps 1 and 2. Note that during this process the tap to the mercury bubbler (1) **must be open** to prevent a pressure build-up in the line.

Manometers to measure pressures in the range 0–760 mmHg need to be connected to various positions as shown on sections **II**, **III** and **IV**. In view of their relatively low cost, mercury U-manometers are the usual choice although, since mercury has a vapour pressure of about 10^{-3} mmHg at room temperature, any line connected to this type of manometer will contain mercury vapour. However, since it is condensable and relatively inert, this can often (but not always!) be ignored. In cases where mercury vapour might cause problems, capacitance manometers or a glass spiral gauge might be used. A simple design of U-manometer is shown in Figure 5.11a, and by incorporating capillary tubing, the amount of mercury required can be minimized. Before use, both legs are first evacuated to high vacuum with the tap open. The tap is then closed to provide a reference vacuum in the sealed leg and the difference in height then gives the line pressure. A dual purpose design [1] incorporating a pressure relief bubbler (Figure 5.11b) is suitable in, for example, section I where pressures may exceed atmospheric upon admission of a gas. Sufficient mercury must be present in the reservoir to prevent the meniscus reaching the bottom of the vertical tube when under vacuum; otherwise air will enter the bubbler and spray mercury into the line.

Condensable gases used on a regular basis can be stored in bulbs grouped together on a manifold (Figure 5.12) which is accessible from the work station.When cooled in liquid nitrogen, the small integral trap on

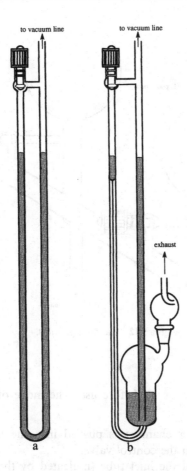

Figure 5.11 Mercury manometer (a) or manometer-bubbler (b) designs.

each bulb enables the complete transfer of a gas into the bulb from the line. This arrangement can also be used for the quantitative transfer of a volatile reagent into a flask attached to the work station provided the volumes of the bulbs and manifold are known (see section 5.3).

Non-condensable gases can either be moved using a Töpler pump, or they can be adsorbed onto a solid support at low temperatures. A Töpler pump uses a mercury column to move gas from the pump chamber into calibrated volumes where its pressure can be measured. Depending on the method used to measure the gas pressure, designs may differ, and Figure 5.13 is based on a description given by Burger and Bercaw in ref. 2. A version using a constant-volume manometer is described in ref. 1. If connected to the line shown in Figure 5.10, the whole pump is evacuated

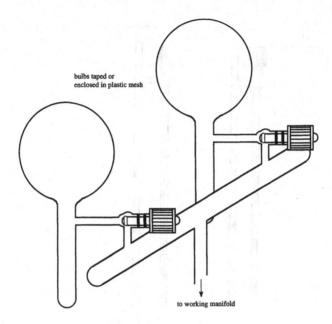

bulbs taped or
enclosed in plastic mesh

to working manifold

Figure 5.12   Gas storage bulbs.

through tap 18 and section **IV** before use. The mode of action is then as follows:

- Mercury in the lower chamber is pushed into the upper chamber by admitting air through the control valve.
- Once past the end of the inlet tube (indicated by the dashed line), the mercury compresses the gas in the upper chamber (V_1) into the calibrated volume (V_2, V_3) where it is trapped by a non-return valve.
- The mercury level is lowered to below the end of the inlet tube by applying vacuum to the control valve.
- This cycle can be automated by using contacts 1 and 2 to switch on and off solenoid valves controlling the air and vacuum supplies.
- Several cycles are necessary to transfer all of the gas (no gas will bubble through the outlet valve on the compression cycle when the process is complete).
- To measure the pressure with this design, the mercury must be raised to beyond the non-return valve (with contact 2 disabled) at which point the inlet tube can be used as a manometer.

The movement of mercury within the pump must be controlled at all times to avoid a sudden rush (a mercury 'hammer') which could break the apparatus and spread mercury around the lab. This somewhat

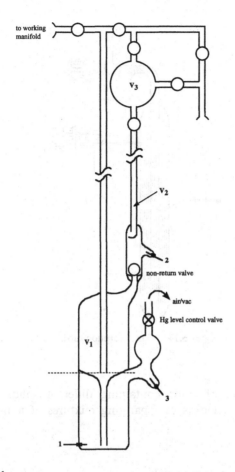

to working
manifold

v_3

v_2

2
non-return valve

air/vac

Hg level control valve

v_1

3

1

Figure 5.13 Töpler pump.

complex operating procedure when coupled with the cost of a Töpler pump and the amount of mercury it must contain makes low temperature adsorption an attractive alternative method for handling non-condensable gases and this technique is described below.

Freshly prepared adsorption tubes (Figure 5.14) containing materials such as molecular sieves or silica gel should be carefully heated to about 300°C under high vacuum before use (the desorbed volatiles can easily carry solid particles into the valves and the vacuum line) and then kept under vacuum. Teflon needle valves are preferred since they give better control over flow rate. Once the trap has been cooled thoroughly to –196°C in liquid nitrogen (which may take some time), the gas can be admitted, the taps closed and the trap moved to another location. By

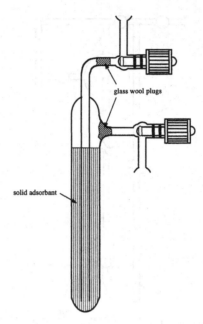

glass wool plugs

solid adsorbant

Figure 5.14   Gas absorption tube.

using a combination of traps containing different solids, preferential adsorption provides a means of separating mixtures of non-condensable gases.

5.3 Manipulating volatile materials

5.3.1 Volatile liquids

Since the manipulation of volatile reagents on a chemical vacuum line relies on their vapour pressure, it is important that non-condensable gases, which slow down transfer rates, are removed. For liquids, this is achieved by a 'freeze–pump–thaw' degassing process as described below.

- Attach the flask containing the volatile compound to the line and evacuate the system up to the tap on the flask.
- Cool the flask to –196°C in liquid nitrogen.
- Open the tap and evacuate the flask.
- Close the tap and allow the liquid to melt, warming as necessary.
- Re-freeze the liquid and open the tap to pump off any non-condensable gases.
- Repeat the process.

Several cycles are usually used to ensure complete removal of gases. With substances that expand on melting (e.g. water and dichloromethane), care is necessary to avoid breaking the container. It is best to use round-bottomed flasks and only fill them to between a quarter and a third full. Tubes can be warmed rapidly to create a film of liquid that allows expansion of the bulk of the frozen solid as it thaws more slowly (this technique is useful when sealing NMR samples).

5.3.2 Condensable gases

By cooling a trap in liquid nitrogen, a condensable gas can be transferred quantitatively from another part of the line, transfer being regarded as complete once the pressure has returned to about 10^{-3} mmHg. With reference to Figure 5.10, to transfer a condensable gas from a storage bulb into a flask connected to **A** or **B**, section **II** (with tap 6 open) and the flask are first evacuated and then tap 7 is closed. The flask is then cooled in liquid nitrogen and the tap on the storage bulb is opened. When all of the gas from the bulb has been condensed into the flask, tap 7 to the vacuum manifold can be opened again to remove any residual non-condensable gases before the tap on the flask is closed. **Beware!** When condensing from a larger to a smaller volume make sure that the resulting pressure at room temperature will not be greater than 1 atm; otherwise there is a risk of explosion (or if greased taps are used these may fly out with some force).

5.3.3 Fractionation trains

The series of traps **C**, **D** and **E** in Figure 5.10 can be used to separate mixtures of volatiles with widely different boiling points. The traps are cooled to temperatures which decrease **C** > **D** > **E** and slush baths (see below) are chosen so that the least volatile component of the mixture is condensed in **C** and successively lower-boiling fractions are collected in **D** and **E**. This method of separation is inefficient unless there is a great difference in volatilities; otherwise more sophisticated methods such as low-temperature distillation or gas chromatography may have to be used. These will not be discussed here, but fairly detailed descriptions can be found in ref. 1.

5.3.4 Slush baths

By cooling different substances to give a mixture of a liquid in equilibrium with its frozen solid, cooling baths with temperatures above −196°C can be prepared (Table 5.1). This is usually done by slowly adding liquid nitrogen or solid CO_2 to the liquid of choice in a clean Dewar flask while stirring the mixture rapidly with a glass rod. Liquid nitrogen that has been kept in a wide-necked Dewar and exposed to the atmosphere for any length of time may have condensed oxygen from the air and should not be used for making up slush baths of combustible liquids. The resulting thick slurry should not have a solid crusty surface which would prevent the immersion of apparatus in the bath. The lifetime of a slush bath can be prolonged by pre-cooling the trap or flask in liquid nitrogen and then allowing it to warm (until samples from the bath no longer freeze on the cooled glassware) before immersion in the bath. After use, allow a slush bath to warm to room temperature before storing the liquid in a suitably labelled bottle for re-use. **Do not** stopper a bottle into which cold residues from a slush bath have been poured. Pressure will build up as the nitrogen comes out of solution.

5.3.5   Quantitative gas measurements

One of the most important features of a high-vacuum line is that it provides the means to measure the amount of a gaseous sample in a particular volume by using the ideal gas law. The volumes of various

Table 5.1 Low-temperature slush baths[a]

Liquid N_2		Solid CO_2	
Solvent	Temp./°C	Solvent	Temp./°C
Diethylene glycol[b]	−10	3-Heptanone	−38
Benzyl alcohol	−15	Cyclohexanone	−46
1,2-Dichlorobenzene	−18	Acetonitrile (tech. grade)	−46
o-Xylene	−29	Chloroform	−61
Chlorobenzene	−45	Acetone	−78
Octane	−56	o- and m-Xylene mixtures[c]	−30 to −70
Chloroform	−63		
Isopropyl acetate	−73		
Ethyl acetate	−84		
Methyl ethyl ketone	−86		
Propan-2-ol[b]	−89		
Heptane	−91		
Hexane	−94		
Toluene	−95		
Methanol	−98		
Ethanol[b]	−116		
Methyl cyclohexane	−126		
Pentane	−131		
2-Methylbutane (isopentane)	−160		

[a]Data taken from: Rondeau, R.E. (1966) *J. Chem. Eng. Data*, **11**, 124; Phipps, A.M. and Hume, D.N. (1968) *J. Chem. Educ.*, **45**, 664.
[b]High-viscosity slush.
[c]Temperature achieved with mixtures of o- and m-xylene have an approximately linear dependence on composition from −26°C (100% o-xylene) to −71°C (100% m-xylene).

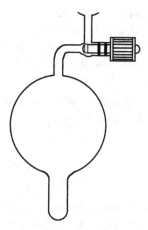

Figure 5.15 Calibrated gas bulb.

sections of the line are determined by measuring the pressure exerted by a known amount of a gas. For this, the volume of a gas bulb (Figure 5.15) is measured as described below.

- Thoroughly clean and dry the bulb.
- Evacuate the bulb on the high-vacuum line.
- Weigh the evacuated bulb.
- Fill the bulb completely with water (mercury can be used for small volumes).
- Re-weigh the filled bulb and note the temperature.
- From the difference in weight and the density of water (or mercury) at ambient temperature, the volume can be calculated.

The calibrated bulb can then be used to introduce a known amount of a condensable gas (CO_2 is a convenient one to use) into various parts of the line using the following procedure.

- Clean and dry the calibrated bulb, attach it to section **II** of the line at **A** (Figure 5.10) and evacuate it *via* tap 7.
- If the required gas is in a storage bulb, open tap 6 and evacuate as far as the tap on the storage bulb. If not, connect a cylinder to **B** and open the tap on the line to evacuate as far as the valve on the cylinder.
- Open tap 8 to the manometer and close tap 7.
- Open the tap on the storage bulb or the cylinder valve and admit the gas to section **II** and the calibrated bulb while monitoring the pressure on the manometer.
- Close the tap on the storage bulb or the cylinder valve at the required pressure.

- Make a note of the pressure and temperature.
- Residual gas in section **II** can be collected in the large cold trap after opening taps 3 and 4 and then tap 5. Tap 5 is then closed and nitrogen admitted through section **I**. The disconnected trap is allowed to warm to ambient temperature in a fume cupboard.

The amount of gas in the calibrated bulb can now be calculated from the known volume, pressure and temperature.

The ideal gas law:

$$PV = nRT$$

where: P is the pressure in atmospheres, V is the volume in litres, n is the number of moles of gas, R is the gas constant, 0.0820575 l atm K^{-1} mol^{-1}, T is the temperature in K.

To calibrate the volume of trap **D** in Figure 5.10, section **II** is evacuated *via* tap 7 and taps 9, 10, 13, 14 and 15 are opened in addition to the tap at **A**. Trap **D** is cooled in liquid nitrogen and the tap on the calibrated bulb is opened to condense the gas into **D**. The trap is allowed to warm to room temperature and the volume of the trap, manometer and connecting tubing is calculated from the known number of moles of gas, the pressure and the temperature. If a mercury manometer is used, the variation of volume with pressure can be plotted on a graph (which should be a straight line) by repeating this process with different amounts of gas. By opening tap 13 during this process, the combined volume including section **III** can be obtained and this can similarly be extended to section **II**.

5.4 High-vacuum pumps

5.4.1 Diffusion pumps

In diffusion pumps, gas molecules are directed towards the outlet by entrainment in a low pressure, high speed vapour stream. The general principle is illustrated in Figure 5.16 which schematically shows a simple metal oil diffusion pump. Vapour from the boiler ascends the chimney, exits the nozzle and is deflected downwards by the cap to produce a vapour jet. Molecules from the vacuum line diffuse into the vapour stream and are driven towards the outlet while the vapour condenses on the wall and is returned to the boiler. A rotary pump provides the initial low pressure required for operation, and diffusion pumps are usually constructed with several nozzles in series to enable operation at higher fore-pressures. Mercury diffusion pumps (Figure 5.17) were once standard

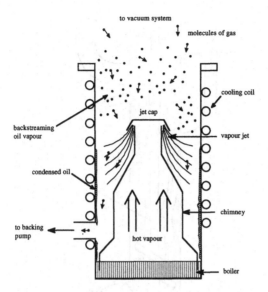

Figure 5.16 Cross-section of an oil-diffusion pump (metal construction).

equipment in research laboratories since they are readily constructed by a skilled glassblower and are relatively inexpensive. However, efforts to reduce the use of mercury because of its toxicity have led to the widespread use of silicone oils in glass diffusion pumps. These have the added advantages that water cooling is often not required and backstreaming of vapour into the vacuum system is not such a problem as it is with mercury pumps. Metal oil diffusion pumps are much more rugged than their glass counterparts and are used extensively in industry. Small versions are available for use on vacuum lines and these can be either water- or air-cooled, the latter being more convenient. Since diffusion pumps require a medium vacuum to operate, they must be by-passed when a large amount of air is pumped through the system. A by-pass eliminates the need to turn off the diffusion pump and then wait for it to heat again when turned back on. It also prevents oxidation of the hot vapour (section 5.2.1). However, this operation is somewhat inconvenient, which is why we fitted a turbomolecular pump to our high-vacuum line described in section 5.5.

5.4.2 Turbomolecular pumps

There is now little difference in price between a small turbomolecular pump and a metal oil diffusion pump of similar capacity complete with its associated fittings. This factor, together with the ease of operation and

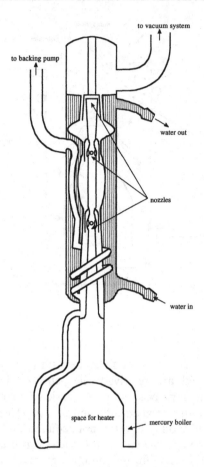

Figure 5.17 Cross-section of a mercury diffusion pump (glass construction).

low maintenance of the turbo pump makes it an attractive source of clean high vacuum. Gas molecules that diffuse into a turbomolecular pump (Figure 5.18) are helped on their way to the pump outlet by a set of blades in a rotor rotating between a fixed set of blades in the stator. Rotation rates of up to 90 000 rev. min^{-1} mean that the blade tips are moving at close to the same velocity as the gas molecules, which collide with the blades and gain momentum towards the outlet as a result (Figure 5.19). Before the turbo pump is turned on, the pressure must first be reduced to about 10^{-2} mmHg, so a backing pump is necessary. Higher operating pressures slow down the rotor and increase the power dissipation of the motor but, providing valves are opened slowly, apparatus can be pumped down from atmospheric pressure without damaging the pump, making a by-pass unnecessary. Power is supplied to the turbo pump from

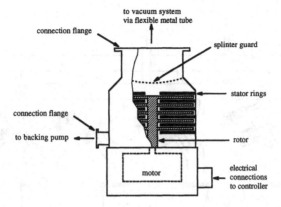

Figure 5.18 Schematic diagram of a turbomolecular pump.

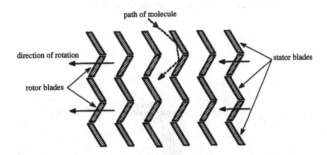

Figure 5.19 Arrangement of blades in a turbomolecular pump.

a remote electronic controller which brings the rotor up to speed gradually and senses any excessive demand on the motor.

5.5 Alternative designs of high-vacuum lines

In laboratories where a permanent multipurpose chemical high-vacuum line is deemed unnecessary, a high-vacuum double manifold may be sufficient for sublimation and vacuum transfer purposes and a greaseless design has been described by Wayda [2]. We have constructed a variation on this design that incorporates a small turbomolecular pump and fits into a 1.5 m fume cupboard and which is very easy to use (Figure 5.20).

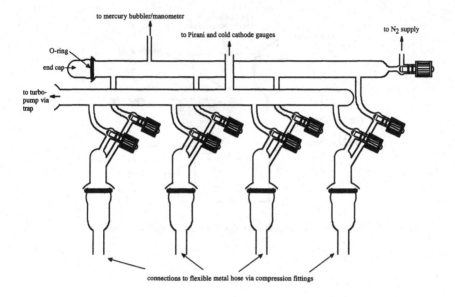

Figure 5.20   Greaseless high-vacuum Schlenk line.

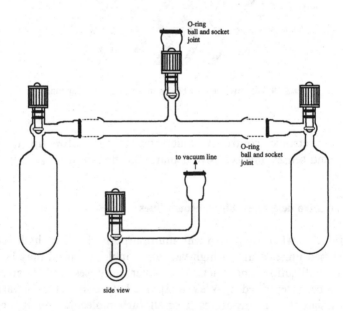

Figure 5.21   Vacuum-transfer adapter for use on a greaseless high-vacuum Schlenk line.

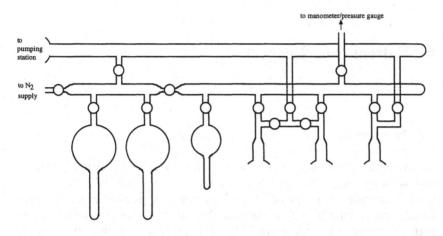

Figure 5.22 Calibrated manifold.

The Teflon screw taps are arranged sideways so that we can see when an O-ring is compressed against the valve seat, thereby reducing the risk of over-tightening the valve. Our workshops modified one end of commercially available screw O-ring compression couplings to fit onto Young's taps so that apparatus can be connected directly to flexible stainless steel hoses. Adapters are used for vacuum transfers (Figure 5.21). Similarly, in the absence of a complete multipurpose high-vacuum line, a self-contained calibrated manifold (Figure 5.22) is often preferred for the quantitative transfer of volatile reagents when Schlenk techniques are being used. The approximate volumes of the various sections of the line are calculated to be convenient for the normal scale of working. These are then calibrated as described in section 5.3 and it is useful to keep a note of volumes and the corresponding number of millimoles per mmHg next to the manifold.

References

1. Shriver, D.F. and Drezdzon, M.A. (1986) *The Manipulation of Air-Sensitive Compounds*, 2nd edn, Wiley, New York.
2. Wayda, A.L. and Darensbourg, M.Y. (eds) (1987) *Experimental Organometallic Chemistry: a Practicum in Synthesis and Characterization*, ACS Symposium Series, Vol. 357.

6 Solvents and Reagents

6.1 Introduction

Good synthetic technique requires due attention to the purity of reagents and solvents. Since repeatability is an important part of experimental work, impurities that may affect the outcome of a reaction must be eliminated, and when choosing a solvent, the air- and moisture-sensitivity of reagents should be considered. Several practical organic texts [1–3] contain useful information and the purification of an extensive range of organic and inorganic compounds is covered in the book by Perrin and Armarego [4]. Detailed descriptions of solvent purification are given in reference 5.

High purity solvent grades cost significantly more than 'reagent grade' and whether you buy an expensive grade or purify a lower grade will usually depend on how frequently and in what volume the solvent is used. Air-stable solid starting materials can often be prepared in reagent grade solvents and then recrystallized from higher purity grades. 'Spectroscopic grade' solvents are free from impurities that absorb in the ultraviolet, and high purity 'HPLC grade' solvents have been distilled and filtered to remove any impurities that may contaminate chromatography columns. Highly isotopically enriched solvents for NMR spectroscopy are normally purchased in small volumes, although these are not always chemically pure, and further drying is often necessary.

6.2 Purifying and storing solvents

6.2.1 General considerations

If you decide not to pay the premium price for a high purity solvent, purification of lower grades usually involves drying, distillation and deoxygenation and can be carried out as a batch or continuous operation. In an average sized research group, it is inconvenient to repeatedly set up and then dismantle the necessary distillation apparatus for solvents used on a regular basis, the preferred option being a set of permanent stills. Important considerations for their design and use are described in section 6.3. Permanent stills are usually run continuously during the working day so that freshly distilled solvents can be collected and used directly from the still. Alternatively, each worker may have their own set of solvent reservoirs, topping them up from the permanent stills when necessary. Once purified, the solvents must be handled under an inert atmosphere using the techniques described in chapters 3 and 5, and, if necessary, stored over a suitable drying agent (usually 3A or 4A molecular sieves). We have permanent stills for diethyl ether, dichloromethane, hexane, tetrahydrofuran, acetonitrile, methanol and toluene, while other solvents such as ethanol, iso-propanol, dimethoxyethane, pyridine, nitromethane and benzonitrile are purified in batches and stored over molecular sieves.

Apart from oxygen and water, other impurities can also interfere with reactions and these are dealt with under the appropriate solvent section. Remember that **most** solvents are **flammable** and/or **toxic** and precautions should be taken to minimize the risk of vapour escape and fire. Wherever possible, permanent stills should be housed in a fume cupboard.

6.2.2 Drying

Azeotropic distillation is a good way to dry many hydrocarbons and halocarbons and is suitable for preliminary purification on a batch basis; otherwise desiccants must be added to the solvent. It is important to choose a desiccant that is compatible with the reactivity of the solvent and, unless the solvent grade has a water content of <0.1%, a preliminary drying stage is essential. After adding the desiccant and allowing the solvent to stand, the solid is separated and the solvent distilled from a more active desiccant. Alternatively, the pre-dried and degassed solvent can be passed down a column of activated molecular sieves or alumina that has been purged with nitrogen. In this case, a small forerun should be discarded or recycled before collecting the rigorously dry solvent. A selection of suitable drying agents and their properties is given below [1–6].

Alumina. Alumina that has been activated by heating to about 200°C (higher temperatures have also been used) in an inert gas stream or *in vacuo* is an efficient desiccant with a maximum capacity of about 10%. This basic material is classed as activity I and is suitable for drying hydrocarbons, chlorinated hydrocarbons, ethers, esters, acetonitrile and pyridine. It also removes peroxides from alkenes and ethers and provides a convenient means of removing the ethanol added to chloroform as a stabilizer. When used in a column, care should be taken not to exceed the capacity of the alumina. Regeneration is possible but if the material has been used to remove peroxides these must first be destroyed by washing with aqueous iron(II) sulphate before heating. Proper procedures must be followed when disposing of used alumina and these are not insignificant when large quantities are involved.

Barium oxide.   Most often used for amines and pyridines and can be used for ethyl acetate. Addition of 5% w/v of anhydrous commercial material can reduce water content to about 30–50 ppm after 24 h. Powdered material that has been exposed to air is likely to be inactive so lumps should be freshly crushed. **Care!** Poisonous dust, avoid inhalation or ingestion.

Boron oxide.   Prepared by heating boric acid to about 300°C, allowing it to cool in a desiccator or under an inert atmosphere and then crushing the solid to give a powder. Can be used for acetone and acetonitrile.

Calcium hydride.   Useful for pre-drying and in stills, this material is best purchased as lumps and crushed prior to use. Suitable for amines, pyridines, HMPA, hydrocarbons, chlorinated hydrocarbons, alcohols, ethers, esters and DMF, but **do not** heat with tetrahydrofuran (reported to react explosively). Can also be used to dry hydrogen, argon, helium and nitrogen gases. Residual solid can be destroyed by careful addition of water (hydrogen evolved, **fire and explosion hazard**).

Lithium aluminium hydride.   Although ethers and hydrocarbons can be dried with this material, the number of reported explosions serve to emphasize the significant risks of using $LiAlH_4$ in solvent stills, especially with tetrahydrofuran. Its use should be **strongly discouraged** in favour of less hazardous desiccants.

Magnesium alkoxides.   Useful for rigorous drying of lighter primary alcohols, especially methanol and ethanol, the alkoxides $Mg(OR)_2$ are usually generated in the solvent still from the pre-dried alcohol and magnesium turnings (about 5–10 g l^{-1}) activated with a few small crystals of iodine. Add only a small amount of alcohol at first, the reaction can be quite vigorous. Should you decide to warm the mixture because the reaction appears sluggish, exercise care – there is sometimes an initiation period! (I have seen an uncontrolled reaction coat the whole of a pristine methanol still with a mixture of magnesium methoxide and magnesium turnings.) Make sure most of the metal has reacted before adding the bulk of the alcohol and then heat slowly to reflux. Only add pre-dried alcohol to the still otherwise insoluble hydrolysis products quickly accumulate and cause the boiling liquid to bump and contaminate the solvent collection reservoir. When this begins to happen, the still should be regenerated.

Molecular sieves. These synthetic crystalline aluminosilicates are available in powder, bead or pellet form with a range of well-defined pore sizes and are the most generally useful and efficient desiccants, types 3A and 4A (with approximate pore sizes of 3 Å and 4 Å, respectively) being most appropriate for drying organic liquids. Most solvents can be conveniently pre-dried over 5% w/v 4A sieves for 12–24 h (methanol, ethanol and acetonitrile with smaller molecular van der Waals radii are adsorbed by 4A sieves, so 3A sieves must be used for these solvents). More efficient drying is achieved by percolating pre-dried, distilled, peroxide-free solvent through a column of sieves. As an illustration, ref. 3 states that the water content of 10 l of a range of ethers, hydrocarbons, chlorinated hydrocarbons, ethyl acetate and acetonitrile can be reduced to better than 20 ppm by passage through a 25 mm × 600 mm column containing 250 g of the appropriate molecular sieve at a rate of $3 \, l \, h^{-1}$. (5A sieves adsorb larger molecules and are used to purify gases, while branched-chain and cyclic compounds are adsorbed by 13X sieves.)

Molecular sieves adsorb atmospheric moisture and fresh sieves should always be generated by heating *in vacuo* or in a stream of dry nitrogen at 300–350°C before use. Used sieves are regenerated without loss of capacity, which can reach a maximum of about 20%. It is usual practice to store dried solvents over molecular sieves, but note that they cause self-condensation in acetone. If molecular sieves are to be used to dry acetone (as recommended in several texts) it is probably best to use a column and check the ^{1}H-NMR of the dried solvent for impurities before use. **Note also** that an explosion involving molecular sieves used to dry 1,1,1-trichloroethane has been reported [8].

Phosphorus(V) oxide. P_4O_{10} is one of the most rapid and efficient desiccants but it has several disadvantages. It is difficult to handle, becoming syrupy as water is taken up, although commercially available materials containing P_4O_{10} on an inert support remain particulate even when hydrated. It is also highly reactive (e.g. towards alcohols, amines and carbonyl compounds) and can cause or catalyse decomposition of solvents (e.g. HMPA, DMSO and acetone). It has been recommended for hydrocarbons, halohydrocarbons, ethers and nitriles, but is less convenient than other desiccants. The supported material is very efficient for drying gases (e.g. hydrogen, oxygen, carbon dioxide, carbon monoxide, sulphur dioxide, nitrogen, methane and ethene) while the oxide itself is used in desiccators. When disposing of residues, always add P_4O_{10} in small portions to ice-water. The acidic solution should then be neutralized and diluted before washing down the sink. (Addition of water to the oxide generates a great deal of heat which could cause the container to crack.) **Care!** Causes burns.

Sodium. The drying capacity of sodium is limited because the crust of solid formed on the surface of the metal inhibits further reaction with water. It should therefore only be used in the form of a fine wire (pressed directly into the solvent) and then only with good quality or pre-dried solvents. It is suitable for hydrocarbons and ethers, although its use is generally discouraged nowadays because of the number of accidents caused when supposedly empty Winchesters containing sodium have been returned to suppliers who then wash them with water. Sodium residues **must** be destroyed in a fume cupboard by the careful addition of ethanol until hydrogen evolution ceases. Let the mixture stand for several hours, stir well to ensure that no coated lumps of sodium remain and then add to a large excess of water and neutralize before washing down the sink. **Caution!** Never add metallic sodium to chlorinated solvents. A vigorous reaction or explosion may result.

Sodium/potassium alloy. Although an insoluble surface coating is not such a problem with the liquid Na–K alloy, there is a significant risk of explosion when this material is handled incorrectly. Since superoxide build-up on the surface can lead to spontaneous ignition, its use in solvent stills should especially be discouraged. Residues should be treated with *tert*-butanol and subsequently with ethanol before being diluted with water and neutralized.

Sodium/benzophenone. Sodium reacts with benzophenone to form an intensely blue ketyl radical that is soluble in ethers, and further reaction with sodium can give the purple radical dianion. These species are very reactive towards oxygen and water to give colourless or yellow products and this self-indicating desiccant is therefore especially convenient for ethereal solvent stills, although by adding small amounts of a polydentate glycol ether such as tetraglyme, it can also be used for hydrocarbons (it is also beneficial to add a small amount of tetraglyme to diethylether stills). It is important that solvents are pre-dried and degassed; otherwise it can be difficult to obtain the blue colour of the ketyl. Freshly cut sodium (a lump measuring 1 cm by 2.5 cm should be sufficient for a 1 l still) is then pressed into the solvent as a fine wire and the pot purged with nitrogen while benzophenone (about 5 g l^{-1}) is added. After swirling to dissolve the benzophenone, a blue colour should form at the metal surface. This localized colour will initially disappear but, on heating to reflux, the bulk of the solvent should gradually turn green and then blue as all of the water and oxidizing impurities are removed. With solvents boiling above 100°C (e.g. toluene, diglyme), prolonged continuous reflux causes decolourization of the ketyl solutions, presumably through decomposition/reduction of benzophenone, but difficulties in obtaining blue or purple colours with other solvents can usually be traced to inefficient pre-drying/degassing or poor still maintenance.

'Titanocene'. The oxygen and moisture sensitivity of the titanium(III) fulvalene species generated by reduction of Cp_2TiCl_2 has led to its use in the purification of solvents [9]. It can be prepared as a green solid [10] and transferred to solvent storage flasks in a dry box before adding hydrocarbon or ether solvents from a still under an inert atmosphere. Pure solvent can then be transferred from the storage flask on a high vacuum line.

6.2.3 Solvents

In addition to the chemical properties of a liquid, its dielectric constant, boiling point, toxicity and cost also need to be considered when determining its suitability as a solvent. A liquid's dielectric constant provides an indication of the types of compounds that are likely to dissolve in it. Non-polar compounds tend to be volatile and soluble in non-polar liquids, whereas polar compounds are usually non-volatile and require polar solvents. The solvents most commonly used for reactions are hydrocarbons (aliphatic and aromatic) and ethers (linear and cyclic), while a wider range finds use during compound isolation and purification. Appendix C lists the properties of a selection of solvents and a few general comments are included below.

Alcohols. The monohydric alcohols methanol, ethanol and propan-2-ol (*iso*propyl alcohol) are probably those most often used as solvents. The reactivity of the hydroxy group restricts use with Lewis acidic halides and amides and oxidizing agents. Preliminary drying over molecular sieves (3A for methanol) followed by distillation from alkoxides formed *in situ* (magnesium for methanol and ethanol, sodium or aluminium for propan-2-ol) gives dry material. Note that methanol is **poisonous** if taken orally (5 cm^3 may prove fatal).

Chlorocarbons. The widespread availability and use of chlorocarbon solvents is decreasing because of worldwide environmental concern, but they are valuable solvents in the research laboratory. Dichloromethane, chloroform and 1,2-dichloroethane are most commonly used although, until recently, 1,1,1-trichloroethane was readily available because of its industrial use as a degreasing agent, and found similar use in the laboratory. **Care:** An explosion involving the detonation of molecular sieves which had been used to dry 1,1,1-trichloroethane (perhaps for a prolonged period) has been described [8]. Activated alumina should therefore be the desiccant of choice for this solvent.

Carbon tetrachloride is carcinogenic and its use as a solvent should be avoided. Dichloromethane is the least reactive and least toxic of these halocarbons. Chloroform reacts slowly with oxygen when exposed to air and light to form poisonous carbonyl dichloride (phosgene). Reagent grade material contains ethanol stabilizer which can be removed by passing the chloroform down an activated alumina column, which also removes water and acidic impurities. Chloroform may also react explosively with acetone. Dichloromethane and 1,2-dichloroethane can be dried by distillation from calcium hydride and then stored over 4A molecular sieves, while activated alumina removes acidic impurities.

Chlorobenzene and o-dichlorobenzene are also useful higher boiling solvents and can be dried over molecular sieves. Again, activated alumina removes traces of acid. **Never** add alkali metals or strong base to chlorinated solvents – this may cause an explosion.

Hydrocarbons. The saturated aliphatic hydrocarbons n-pentane and n-hexane are the standard non-polar hydrocarbon solvents of choice for continuous stills while higher straight-chain and cyclic homologues such as heptane, octane, cyclohexane and methylcyclohexane are useful when higher temperatures are required. Commercial pentane, hexane, cyclohexane, etc. are mixtures of alkanes that are free of aromatics but which are often contaminated by alkenes. The pure single compounds (which are often not necessary) are more expensive and it is wise to check the catalogue carefully before making a purchase. Alkenes can be removed from the commercial mixtures by repeatedly shaking with concentrated sulphuric acid (5% by volume) and then washing with water until neutral prior to any drying procedures. Of the aromatic solvents, benzene is carcinogenic, leaving toluene as the usual choice for routine use nowadays, while xylenes and mesitylene are used for higher temperature applications. Mixtures of aliphatic hydrocarbons containing small amounts of aromatics are sold under the confusing name of petroleum ether and are supplied as fractions with 20°C boiling ranges (e.g. 40–60, 60–80, etc.). Hydrocarbons are easier to dry than more polar solvents, and alumina, calcium hydride and 4A sieves are all convenient desiccants. For continuous stills, calcium hydride or self-indicating sodium/benzophenone (with the addition of 10% tetraglyme to solubilize the ketyl) can be used, although with toluene the latter tends to darken significantly on prolonged reflux. All of these solvents have some toxic effects and it is a good general policy to avoid breathing their vapours.

Esters. Ethyl acetate is sometimes used for recrystallization of polar compounds and can be dried over potassium carbonate, P_4O_{10} or 4A sieves and then distilled, or by refluxing over and then distillation from barium oxide (5 g l^{-1}). It can be stored over 4A molecular sieves.

Ethers. Along with the hydrocarbons, this is the other most common class of solvents. Most synthetic metalorganic laboratories have diethyl ether and tetrahydrofuran (THF) in continuous stills. Dibutyl or diphenyl ethers are useful when diethyl ether is too volatile, and the polyethers or glymes, e.g. dimethoxyethane (DME) and diglyme, are often suitable solvents for compounds of the Group 1 and Group 2 metals (hence the use of tetraglyme with sodium-benzophenone in solvent stills). Dioxane is another cyclic ether that finds use as a solvent and, like tetrahydrofuran, it often forms complexes with Lewis acidic metal centres. Ethers are prone to peroxide formation (especially diisopropylether), and this is accelerated by air, heat and light. Evaporation or distillation to near dryness of an ether containing peroxides may cause a serious explosion.

To test for peroxides, add 1 cm^3 of the solvent to 1 cm^3 of a 10% solution of sodium iodide in acetic acid. No yellow colour (low concentrations of peroxide) should be present after 1 min. If a brown colour develops then high concentrations of peroxides are present. Distillation from sodium-benzophenone destroys peroxides, as does shaking the ether with acidified aqueous iron(II) sulphate. Alternatively, they can be removed on an activated alumina column (section 6.2.2).

Purified ethers should be stored under nitrogen in dark containers, preferably in a cool place, and not for more than a few weeks. Commercial ethers may contain small amounts of 2,6-di-*tert*-butyl-4-methylphenol (BHT) as a stabilizer. An ether worth considering as a solvent because of its reluctance to form peroxides (see below) and its convenient boiling point (55–56°C) is *tert*-butyl methyl ether. Produced on a large scale as a high-octane fuel additive, it is increasingly being used in organic chemistry, although it has yet to gain significant popularity in metalorganic laboratories.

Ketones. Acetone is susceptible to acid and base catalysed condensation and this should be borne in mind before using it as a solvent for reactions or for recrystallization. This sensitivity towards a range of desiccants makes acetone difficult to dry, although overnight storage over 10% w/v 3A sieves (significant condensation occurs over longer periods of time) followed by stirring over boron oxide (5% w/v) for 24 h and then distillation has been recommended [11]. Very pure acetone can be obtained from the sodium iodide addition compound as follows [5]. Acetone saturated with dry sodium iodide at 25–30°C is decanted from remaining solid and cooled to –10°C to produce crystals of the adduct which are filtered, transferred to a flask and warmed to above 26°C. The resulting liquid is removed and distilled with rejection of the last 10%.

Nitriles. These compounds often bind to transition metals, but when a polar medium is required (e.g. for salts containing organic cations) acetonitrile or the higher boiling nitriles are useful solvents. They are less easy to purify and dry than hydrocarbons and ethers, and acetonitrile may contain ammonia and acetic acid (which are degradation products). Strongly acidic, basic or reducing desiccants cannot be used, but reflux over and distillation from calcium hydride provides material which is sufficiently pure for most purposes, and subsequent passage down a column of highly activated alumina provides very dry material. Boron oxide and P_4O_{10} are other possible desiccants for acetonitrile, although the latter causes the formation of a yellow gel in the still-pot.

Water. Even when the majority of your work involves moisture-sensitive compounds there are likely to be occasions, usually during the preparation of starting materials, when purified water is required as the solvent. Commercially available deionizing systems remove dissolved salts, whereas triple distillation with the incorporation of a potassium permanganate stage removes organic and inorganic impurities. Deoxygenation is achieved using standard degassing techniques.

Other solvents. A range of other polar, aprotic solvents are available for ionic compounds. Dimethylsulphoxide (DMSO) is a versatile solvent for inorganic and organic compounds including polymers. Dry over 4A sieves and then distil under reduced pressure (75°C/12 mmHg or preferably at a lower pressure), discarding the first 20%, and store over 4A sieves. Its thermal decomposition is acid catalysed. DMSO readily penetrates the skin, providing a vehicle into the body fluids for any toxic solutes that happen to be around, so **avoid all contact with the skin**. Sulpholane, another oxo-sulphur compound, has a moderately high dielectric constant and is stable up to 220°C. It has been used as a solvent for organic fluoro-compounds and for polymers, but one drawback is the difficulty of removing it from reaction mixtures due to its involatility. Dry over KOH overnight (500 g l^{-1}) and distil under high vacuum. Note that it is a solid at room temperature (m.p. 28.5°C).

N,N-Dimethylformamide (DMF) is a polar solvent with a large liquid range but it is not easy to purify and is susceptible to acid- and base-catalysed decomposition. Dry over three sequential batches of 3A sieves (5% w/v, 24 h), or alternatively, overnight over calcium hydride or P_4O_{10} then distil (under reduced pressure) onto 3A sieves. DMF is absorbed through the skin.

Nitromethane exists in equilibrium with its tautomeric aci-form, a pseudo-acid, and as such is quite reactive. It can be dried by standing over calcium chloride, filtering and distilling onto molecular sieves. Passage of distilled material down three activated alumina columns (35 cm × 2.5 cm each for 1 l) has been reported to reduce the water concentration to 0.003 M. P_4O_{10} should not be used to dry nitromethane.

Note that nitrobenzene is **toxic** and readily absorbed through the skin.

Pyridine has been used extensively as a solvent. It is hygroscopic and can be dried over 4A sieves, by passing down an activated alumina column or by distillation from calcium hydride or barium oxide onto 4A sieves. It is **toxic** and is absorbed through the skin.

6.3 Design and maintenance of solvent stills

Continuous solvent stills are simply a means of separating a solvent from a solid desiccant and as such are not designed to have a high distillation efficiency. They usually consist of a distillation pot, a condenser and a reservoir, arranged vertically to economize on space and allow distillate to be either collected or returned to the pot, and have a connection to an inert gas supply. Note that this setup does not provide for fractionation and if you suspect that volatile impurities may be present in a solvent, they should be removed in separate purification steps (unless they react with the desiccant).

6.3.1 Design considerations

Stills must be sturdy and allow easy operation, and range from those comprising readily available items of glassware (e.g. Figure 6.1a) [1] to customized versions with integrated cooling (Figure 6.1b). The all-in-one designs have the advantage that there are fewer joints which might leak and allow in air. The more elaborate still-head shown in Figure 6.2 is a modified version of a commercially available design described by Wayda and Bianconi [12]. In all of these designs, the solvent vapour rises, condenses and runs into a reservoir where it collects if the tap is closed. Under continuous distillation, when the reservoir is full the solvent overflows back into the still-pot *via* the vapour inlet. The volume of the reservoir up to the vapour inlet should be smaller than the volume of the still-pot so that with the pot at least one-third to one-half full, continuous distillation will not result in it boiling to dryness. Opening the tap drains the reservoir into the pot.

An access port above the level of the vapour inlet provides for the removal of purified solvent with a syringe. This port should be fitted with a septum cap after opening the tap or removing the screw cap, and then purged with nitrogen by inserting a bleed needle before taking out any solvent. The tube extending to the bottom of the reservoir in Figure 6.1b and the direct connection to the nitrogen inlet helps to prevent the nitrogen purge carrying hot solvent vapour out through the syringe port where it could attack the rubber septum cap. The all O-ring design in

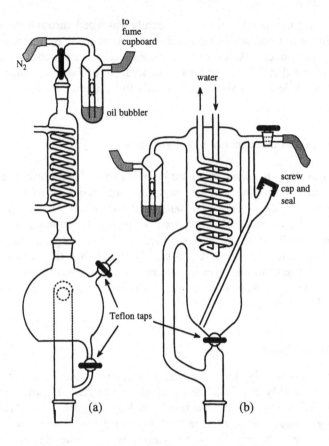

Figure 6.1 Still-head designs for continuous distillation.

Figure 6.2 incorporates a Teflon 'Flickit' valve so that straight-through access is possible with only a half-turn of the valve. Also shown in Figure 6.2 is the arrangement necessary to transfer solvent directly into a receiver. A nearby connection to vacuum must be available so that the connection to the flask can be evacuated and purged with nitrogen prior to transfer.

Safety note: The still-heads shown in Figures 6.1 and 6.2 have taps for the connection to the nitrogen supply. During operation, these taps **must be open** at all times to prevent a pressure build-up and possible explosion. In many departments, the incorporation of such taps is now forbidden because of explosion incidents caused by the failure of researchers to check that the tap is open before turning on the still.

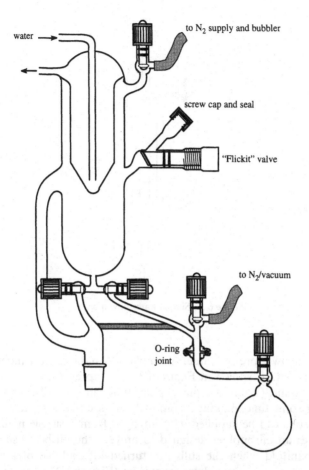

water →

to N₂ supply and bubbler

screw cap and seal

"Flickit" valve

to N₂/vacuum

O-ring joint

Figure 6.2 A Wayda/Bianconi still-head.

6.3.2 Maintenance and use

It is usual in a lab where anhydrous, oxygen-free solvents are required on a routine basis to have several continuous stills set up permanently. It is therefore important that all workers in the lab know how to use and maintain the stills properly so that dismantling and regeneration can be minimized and everyone can be confident that the solvents are pure. Where stills are set up with their water cooling in series, those containing the most volatile solvents (e.g. diethyl ether, dichloromethane) should be closest to the water supply where the water will be coldest. Some means of monitoring the flow rate should be incorporated, preferably with a cut-off switch so that power to the heating mantles is turned off when the flow

water from stills

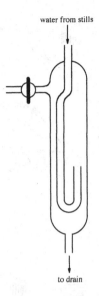

to drain

Figure 6.3 Simple water-flow monitor.

rate drops below a pre-set value. A simple flow indicator for incorporation after the last still is shown in Figure 6.3. The tap enables it to be drained by venting to air. (We found that a plain tube tends to fill completely with water after some time, making it impossible to see the flow rate.)

Several stills can be supplied with nitrogen from a simple manifold that incorporates an oil bubbler designed to prevent the oil being sucked back into the manifold when the stills are turned off and the pressure in the system drops as the gas volume contracts (Figure 6.4). A slow nitrogen flow rate should be maintained through the bubbler at all times, and the nitrogen flow should be increased when solvent is taken from a still or when the heating is turned off. The water flow meter and the oil bubbler should be clearly visible from most parts of the lab to enable routine monitoring when the stills are on.

Deciding on the volume of the still-pot is something of a compromise. Smaller pots have to be refilled more often and this usually means that they also need more frequent regeneration. On the other hand, stills containing much larger volumes of solvents take up more space, present significant toxicity and fire hazards and are more difficult to regenerate should they become contaminated. We find that 1-l flasks don't create too much of a maintenance burden provided everyone fulfils their obligations on the solvent rota.

Still-pots are most often heated with heating mantles, although it is also possible to use infrared lamps. To prevent overheating (section 7.5), the

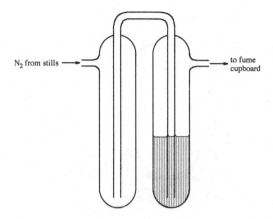

N₂ from stills → … → to fume cupboard

Figure 6.4 Oil bubbler.

level of solvent should not be allowed to drop below the top of the mantle and the still-pot should **never** be allowed to boil dry. A neat solution to the overheating problem is to incorporate a smaller round-bottomed flask (e.g. 250 ml) into the still-pot (Figure 6.5) [13]. A smaller heating mantle can then be used and all but about 120 ml of the contents can be distilled without overheating the walls of the flask. Apart from the additional advantage of reducing the outlay on heating mantles, for the more volatile solvents it is easier to control the heating with a smaller mantle, thereby reducing the chances of overloading the condenser and getting solvent in the nitrogen line.

Teflon sleeves should be used on ground-glass joints used to connect the still-head to the distillation pot since hot solvent can leach out the grease from greased joints, contaminating the solvent and causing the joints to stick. For the same reason, Teflon taps or screw valves should be used on still-heads.

When the sodium/benzophenone in a still loses its purple/blue colour, the colour can often be regenerated by adding more sodium and/or benzophenone under a nitrogen stream once the still has cooled. After some time this will become impractical because of the build-up of solid decomposition products in the pot and the still must then be dismantled for regeneration. Distil most of the remaining solvent into the reservoir before connecting a clean, dried flask containing some sodium wire and benzophenone to the still-head. The solvent from the reservoir can then be run into the new still-pot, additional pre-dried solvent added as necessary and the still turned on. Non-self-indicating stills should also be regenerated periodically or when they start to behave erratically (e.g. when a methanol still begins to bump badly).

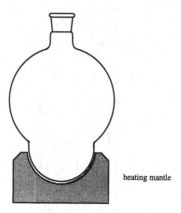

heating mantle

Figure 6.5 Still-pot modified for use with a smaller heating mantle.

Smaller portable versions of the still-head shown in Figure 6.1b are very useful for distilling less frequently used solvents. Such solvents can be stored in a round-bottomed flask over the appropriate desiccant and, when necessary, the clean, dry still-head is pumped out (after attaching a small flask to the joint), flushed with nitrogen and attached to the solvent flask for distillation. After use the still is dismantled and the still-head dried in the oven. Small still-heads of this type are also ideal for cracking and distilling cyclopentadiene immediately prior to use.

Additional safety points

1. It is a good idea to put a spill tray containing an absorbent material (e.g. sand) under the heating mantles.
2. An electronic flow meter incorporated into the water supply will turn off the power to the heating mantles when the flow rate falls below a pre-set level.
3. If the stills are housed in a cabinet or fume cupboard, heat detectors can be positioned so as to activate dry-powder fire extinguishers, turn off the power to the heating mantles and shut off the water supply (by means of a solenoid valve) in the event of a fire. Note that a fire in a fume cupboard is rapidly spread by being drawn through the ventilation ducting by the extraction fans.

Because of the hazards associated with solvent stills, some groups now prefer to dry their solvents by passing them through stainless steel cylinders packed with molecular sieves.

6.4   Reagents

This section is included to give some idea of the types of compound that can be used to effect commonly encountered transformations in metalorganic chemistry, although it is by no means intended to be an exhaustive list.

6.4.1   Halogenation

The replacement of halide with other ligands is a fundamental synthetic step (see following section) so in many cases it will be necessary to prepare a suitable starting halide using a halogenating reagent. Unless your research specifically concerns fluoro complexes, it is unlikely that you will need to prepare metal fluorides. Fluorine and hydrogen fluoride require special handling techniques [6,14] because of their reactivity and toxicity, although lower oxidation state metal fluorides may be obtained by halide exchange reactions.

The halogens. Cl_2, Br_2 or I_2 may be suitable reagents when oxidation of the metal centre is possible, as in the preparation of anhydrous metal halides from metals or oxidative addition to lower oxidation-state metal complexes. They can be used in the elemental form as described in chapter 9 or as solutions in an appropriate solvent.

Hydrogen halides.   Gaseous HCl and HBr can also be used for the conversion of metals to anhydrous halides, where the product is often a lower oxidation-state metal halide than that obtained from the metal and halogen. Halide ligands can sometimes be introduced by the protonation of e.g. dialkylamido and alkyl groups by HX with loss of amine or alkane, respectively.

Halides of main group elements.   Metal oxides can be converted to halides or oxyhalides with $SOCl_2$, PCl_5, CCl_4, $COBr_2$. Of these, thionyl chloride is perhaps the most convenient, especially for large-scale preparations, since it is a liquid and can be used as the solvent. Early transition metal oxyhalides such as $VOCl_3$, $MoOCl_4$ and $WOCl_4$ are readily prepared by this route. Carbon tetrachloride is usually entrained in nitrogen and passed over the heated metal oxide, but care is needed since CCl_4 and the $COCl_2$ (phosgene) produced in the reaction are toxic. The use of carbonyl bromide $COBr_2$ to prepare metal oxybromides (with reduction in some cases) has been reported [15], although this reagent is not readily available. On a smaller scale, oxalyl halides $C_2O_2X_2$ (X = Cl, Br) can act as a halide source in reactions of metal oxo compounds, but reduction of the metal centre may occur. Acyl halides CH_3COX (X = F, Cl, Br, I) have been used to convert metal dialkylamides and alkoxides to halides, and it has been proposed that these reactions are catalysed by traces of HX present in the acyl halides. Trimethylsilyl chloride is useful for the conversion of metal carboxylate complexes to the halide derivatives, with the elimination of the trimethylsilyl ester.

6.4.2 Ligand metathesis/exchange

The replacement of one type of ligand for another is perhaps the most fundamental of reactions at the metal centre. In addition to the exchange of an atom or a group of atoms between the metal centre and another compound (metathesis), I have mentioned a few reagents that can be used to remove ligands from the metal and thereby facilitate ligand substitution reactions.

Metal salts of anionic ligands. The insolubility of main group metal halides is often the driving force in the exchange of halide ligands for a wide range of anionic groups, including other halides. For example alkoxo, carboxylato, cyclopentadienyl, dialkylamido, and thiolato complexes can be prepared by reacting the metal halide with an appropriate metal salt (most often of an alkali metal) obtained by deprotonation of the neutral molecule (alcohol, carboxylic acid, etc.) with the metal itself or with a metal hydride or alkyl compound. Alkylation reactions are described in section 6.4.3. In most cases, the solvent of choice will be an ether, although due consideration must be given to the relative solubilities of the starting materials and products to enable the metal halide produced to be separated from the soluble products by filtration. For example, in the preparation of sparingly soluble transition metal methoxides, the starting metal halide can be reacted with lithium methoxide in methanol, since lithium chloride is somewhat soluble in this solvent and can therefore be removed by washing with methanol.

Trimethylsilyl compounds. Strong silicon–oxygen and silicon–halogen (especially Si–F) bonds enable the introduction of a variety of ligands X in reactions between metal halide or oxo complexes and Me_3SiX, the volatile Me_3SiF, Me_3SiCl or $Me_3SiOSiMe_3$ being readily removed from the reaction product. For example, highly chlorinated compounds react with $Me_3SiOSiMe_3$ to give oxide derivatives, metal amides are prepared from silylamines Me_3SiNRR', and silylated cyclopentadienes $C_5R_5SiMe_3$ have been used to prepare cyclopentadienyl complexes (the analogous tin reagents might also be considered for this type of conversion). Tin analogues are also used, but note that Me_3SnCl is **very toxic** and volatile.

Main group hydrides. Hydrides such as R_3SnH, AlH_4^-, R_3BH^-, R_3AlH^-, $(RO)_3BH^-$, or $(RO)_3AlH^-$ can be used in the preparation of hydride complexes from halides, although reduction is also a possibility (see section 6.4.4).

Other reagents. Halide abstraction with a soluble silver (e.g. $AgBF_4$) or thallium salt (**very toxic**, use proper disposal procedures) provides a means of substituting halide by neutral ligands, where precipitation of the insoluble halide drives the reaction to completion. Trimethylamine *N*-oxide is a useful reagent for labilization of carbonyl ligands under mild conditions, as an alternative to thermal or photochemical methods.

6.4.3 Alkylation

The most common alkylating agents are the alkyls of the elements Li, Mg, Zn, Al, although the heterometallic lithium copper alkylates $LiCuR_2$ have also been used. It is worth noting that in reactions involving aluminium alkyls, only one alkyl group per aluminium is usually transferred.

Organolithium and Grignard reagents. A wide range of these reagents is available commercially, but in many cases lithium alkyls and Grignard reagents are prepared as required by reaction of the metal with an alkyl halide in a hydrocarbon solvent (it has been suggested that cyclohexane is the best solvent for alkyl lithium compounds [2]). These reactions are accelerated by sonication in an ultrasonic bath or reactor (section 12.6), and when preparing organolithium compounds, the lithium used should contain 2% sodium. Ethers are suitable solvents for Grignard reagents, but are more reactive towards organolithium compounds. The concentration of the filtered solution should be determined by titration, and it is also good practice to check the concentration of commercial solutions. In the Gilman 'double titration', a sample of the solution is hydrolysed with distilled water and then titrated with dilute acid using phenolphthalein indicator to give the total amount of base. A second sample is then treated with an excess of an alkyl halide and the hydrolysis and titration repeated. The difference in the titres is then used to calculate the concentration of metal alkyl.

An alternative procedure [16] uses the formation of an intensely coloured organic dianion as the end point in a direct titration with the metal alkyl, and suitable indicators for organolithium reagents include diphenyl acetic acid, *N*-pivaloyl-*o*-toluidine, *N*-pivaloyl-*o*-benzylaniline or 1,3-diphenyl-2-propanone *p*-toluenesulphonylhydrazone. Typically, 0.9–2.0 mmol of the indicator is dissolved in dry THF (5–10 cm^3) in a 25 cm^3 flask under nitrogen or argon. The organolithium solution is added to the rapidly stirred THF solution from a gas-tight 1 cm^3 syringe through a septum cap. The end point (i.e. the appearance of a permanent colour) represents the addition of one equivalent of metal alkyl.

A second alternative to the Gilman method uses the disappearance of an intense colour due to complex formation between the metal alkyl and a small amount of an indicator upon titration with a standard solution of acid or alcohol [17]. In this case, a small measured volume of the metal alkyl solution is added to a dry solution of the indicator, a heterocyclic compound with 1,4-nitrogen donor atoms (e.g. *o*-phenanthroline, 2,2'-bipyridine, 2,2'-diquinolinyl) under nitrogen or argon. A standard solution of *sec*-butanol in toluene is then added until the colour is discharged. This method is quite versatile in that it can be used for Grignard reagents and amide bases (section 6.4.6).

These highly reactive alkyls, especially the aluminium and zinc compounds, require careful handling under inert atmosphere conditions. Pure aluminium and zinc alkyls are supplied in metal cylinders fitted with valves, and while metal alkyl solutions are packaged under argon in 'Sure-Seal' bottles that provide some protection from the atmosphere, it is better to transfer the fresh solution to storage vessels of 50–200 cm^3 capacity fitted with Teflon screw taps and keep these in the refrigerator. The design shown in Figure 6.6 with a built-in glass collar and a tapered bottom will stand on a flat surface without support and make the removal of the last few cm^3 of solution somewhat easier. The optional ground-glass tap is an extra insurance against the greaseless valve developing a leak when cooled.

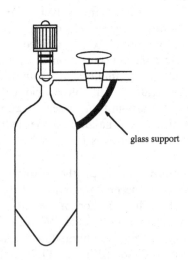

glass support

Figure 6.6   Storage vessel for air-sensitive liquids.

6.4.4   Reduction

The choice of reducing agent can be critical in the preparation of lower oxidation state metal compounds, and it is important that you are aware of the range available. Chemical reductants and oxidants have been reviewed [18].

Metals.   Metals such as Li, Na, Mg, Al and Zn are often used in excess as powders. Very finely divided metals are especially effective (Riecke metals), and the various methods available for the preparation of highly activated metals has been reviewed along with some of their applications [19].

Metal amalgams.   The careful addition of sodium to clean mercury is exothermic and produces an amalgam which can be liquid or solid depending on the percentage of sodium. A convenient liquid amalgam contains 0.4% Na and ether solvents can be used for sodium amalgam reductions. I have also used 0.4% amalgam for reductions in dichloromethane without any problems (but note that **halogenated solvents** should **never be used** with **metallic** sodium). Granular zinc can be amalgamated by treatment with a solution of a mercury(II) salt.

Radical anions.   The addition of catalytic or stoichiometric amounts of electron transfer agents such as benzophenone or naphthalene to the alkali metals produces intensely coloured (blue or green, respectively) radical anions which act as reducing agents. Standardized solutions of these reagents can be prepared.

Sonication of magnesium with anthracene in tetrahydrofuran gives the highly reactive metallated complex $[Mg(C_{12}H_{10})(THF)_3]$, a stabilized radical dianion that can be isolated as a thermally unstable solid and used as a reducing agent.

Potassium-graphite. C_8K, a bronze coloured solid prepared by heating potassium with graphite at 150°C under argon, is a versatile reducing agent and has been used in the preparation of the Ti(II) and Nb(II) complexes $[TiCl_2(py)_4]$ and $[NbCl_2(py)_4]$ [20]. Such highly reduced transition metal complexes are themselves useful reducing agents, and Ti(III) compounds may be used in reactions involving oxygen abstraction.

Lithium nitride. Li_3N has been used to reduce some metal halides, e.g. in the preparation of Ti(III) compounds [21], but note that there is a risk of explosion with this material which therefore requires careful handling.

Main group hydrides.   These compounds were mentioned in section 6.4.2 and are useful reducing agents, although their behaviour is not always predictable (for example, BH_4^- can also act as a chelating ligand) and the monohydrido reagents may offer better stoichiometric control. Solutions of boron or aluminium hydrides can be standardized by hydrolysing a known volume with dilute acid and measuring the amount of hydrogen evolved using a gas burette as described in section 7.2.3. Hydroxylamine can also be used as a reducing agent, e.g. in the preparation of $[Pd(PPh_3)_4]$ [22].

Other reducing agents. H_2 and CO are used at elevated temperatures for the reduction of solid oxides while in solution, trialkyl phosphines can abstract oxygen or halogen from higher oxidation state complexes.

Electrochemical techniques. If the apparatus is available, electrochemical methods offer an alternative method of reduction, once the reduction potential has been determined by cyclic voltammetry. The degree of reduction can be controlled by measuring the current and stopping the electrolysis after transfer of the requisite amount of charge. See section 12.2 for further details.

6.4.5 Oxidation

Oxidation of a metal centre is less frequently encountered than reduction as a synthetic step in metalorganic chemistry and there are fewer well defined reagents for stoichiometric reactions. When using organic solvents, the possibility of a vigorous reaction with the oxidizing reagent should always be borne in mind (e.g. peroxide formation with ethers). See ref. 18 for a review of chemical oxidants and reductants.

Oxygen and the halogens. Oxidation can sometimes be achieved simply by bubbling air or oxygen through a solution of the compound, but other reagents offer greater control. Standard solutions of the halogens are readily prepared for stoichiometric oxidative additions (first check that the solvent will not react).

Metal compounds. Ag(I) salts are useful as one-electron oxidants and produce metallic silver, while the ferrocene produced in oxidations with the ferricinium ion Cp_2Fe^+ can be washed out with non-polar solvents or even sublimed. Compounds of Ce(IV) or the higher oxidation state transition metals might also be considered, but reduction products may be more difficult to separate during work-up.

Oxide transfer and related reagents. These include hydrogen peroxide, *tert*-butyl hydroperoxide, nitrous oxide, iodosyl benzene (PhIO), amine *N*-oxides and in some cases phosphine oxides. These should never be used in ether solvents (including THF) because of the possibility of forming explosive peroxides.

Organic azides. RN_3, especially the more stable aryl compounds, will oxidatively add to a range of lower oxidation state metals to give organoimido complexes (it is possible that with trimethylsilyl azide elimination of trimethylsilyl halide will result in a nitrido rather than organoimido product).

Electrochemical techniques. As in the case of reduction, if the oxidation potential is known, electrochemical methods offer an alternative method of controlled oxidation (section 12.2).

6.4.6 Deprotonation

The type of base required for a given deprotonation will depend on the pK_a of the substrate. In metalorganic chemistry, bases are most frequently used in the generation of anions for ligand metathesis or in the deprotonation of coordinated ligands. The organic chemistry involved in particular ligand syntheses may require very specific bases, and the properties and relative merits of a wide selection have been described [2].

Amines. Amines such as pyridine and triethylamine are most often used as acid scavengers although they can also act as catalysts in proton transfer reactions. Note that ammonia is used in the preparation of metal alkoxides. Amine basicity is enhanced when the proton can be sandwiched between two tertiary amine groups as, for example, in 1,8-bis(dimethylamino)naphthalene which is sold as 'Proton Sponge'. Other non-nucleophilic amine bases are available commercially, e.g. Dabco (1,4-diazabicyclo[2.2.2]octane), DBN (1,5-diazabicyclo[4.3.0]non-5-ene) and DBU (1,8-diazabicyclo[5.4.0]undec-7-ene), from Aldrich.

Metal amides. Usually produced by reaction of the amine with *n*-butyllithium, these strong bases are soluble in ethers and the bulky, non-nucleophilic lithium diisopropylamide (LDA) is commonly used in synthetic organic chemistry. Sodium bis(trimethylsilyl)amide $NaN(SiMe_3)_2$ is conveniently prepared by heating hexamethyldisilazane with sodium hydride (15% excess) in toluene under reflux for at least 24 h. The solution obtained by filtering off unreacted sodium hydride is stable and can be stored for extended periods of time or alternatively the solid amide can be obtained by stripping the solvent. Sodamide is prepared by adding sodium to liquid ammonia (section 12.8) followed by addition of a catalytic amount (a few crystals) of iron(III) sulphate or other iron(III) salt to discharge the blue colour of the solution.

Lithium alkyls. *n*-Butyllithium finds widespread use as a strong base and is available commercially in hexane solution in concentrations ranging from 1.0 to 2.5 M, but avoid 10 M solutions since their viscosity makes them difficult to handle by syringe techniques and thereby unsuitable for general use. Rather than use multiple syringe transfers, large volumes of *n*-butyllithium solution should be measured *via* cannula into the reaction flask from a measuring cylinder fitted with a screw tap (Figure 7.3). Organolithium reagents are more reactive when their oligomeric structures are dissociated by the addition of chelating ligands such as tetramethylethylenediamine (TMEDA) or crown ethers. To prevent attack on solvents such as THF by these activated reagents, reactions should be carried out below −70°C.

Alkali metal alkoxides. A wide range of alkali metal alkoxides with differing solubilities is readily prepared from the alcohol and either the metal, metal hydride (for sodium and potassium) or *n*-butyllithium, the bulkier alkoxides being more soluble in hydrocarbon solvents. It is best to sublime potassium and lithium *tert*-butoxides under high vacuum to ensure complete removal of residual alcohol. Note that addition of sodium or potassium *tert*-butoxides to an alkyllithium or dialkylamide enhances the reactivity of these bases.

Metals and metal hydrides. As already mentioned, alkali metals and metal hydrides are useful for the generation of anionic reagents, (e.g. cyclopentadienide or alkoxides). Sodium and potassium hydrides are conveniently supplied as suspensions in mineral oil that can be transferred to a tared flask in air before washing off the oil with hexanes under nitrogen, drying the finely divided solid *in vacuo* and re-weighing.

Other non-aqueous bases. Quaternary ammonium hydroxides are available as solutions in methanol although we have found residual halide to be present in some cases. Phosphorus ylides have also occasionally been used for proton abstraction.

6.5 Gases

Gaseous reagents can be more problematic than liquids, especially if you don't have access to a calibrated vacuum manifold. However, if an excess of the gas is required, it can simply be bubbled through the reaction mixture, provided a fume cupboard is used and suitable measures are taken to absorb/destroy any hazardous gases. The gas can be taken from a cylinder or generated by a suitable reaction, depending on the purity required and on the scale and frequency of use. Safety data sheets for all gases in the laboratory should be readily accessible (these should be provided by the company that supplied the gas). Consult these data sheets before using any gas.

In addition to the hazards due to the chemical properties of gases, you must be fully aware of the **risks** associated with handling cylinders of compressed gases, often at very high pressures.

6.5.1 Gas cylinders

A wide range of gases is commercially available in different types of cylinder depending on the properties of the gas. You will probably be most familiar with the 'tall' drawn steel cylinders used for high pressure gases such as nitrogen and argon, which have dimensions of about 1.5 m long × 23 cm diameter, but sizes also range down to the small portable lecture bottles (typically 37 × 5 cm). High pressure cylinders contain gases at up to 170–200 atm (about 2500 lb inch^{-2}) and as such represent a considerable amount of stored energy (1 atm = 1.01 bar = 101 kPa = 14.7 lb inch^{-2} = 760 mmHg = 407.14 inch of H_2O). A sudden release of pressure could propel the cylinder like a missile through the laboratory with disastrous consequences, so take appropriate care when handling or moving these containers, especially the larger ones.

- always bear in mind the mass of the cylinder
- observe all manual handling regulations
- beware of trapping fingers between the cylinder and other objects
- use a cylinder trolley (preferably one with an extra wheel to support the weight when the cylinder is tilted) and ensure that the cylinder is properly secured
- fit valve protection caps or guards when the cylinder is moved or when it is not in use
- stand the cylinder upright and secure it properly to prevent it from toppling over

Never leave a high pressure cylinder unsupported.

Gases such as ammonia which have higher boiling points are supplied in lighter, squat-form, welded steel or aluminium cylinders (typical sizes are 38 × 27 cm or 91 × 33 cm) under lower pressures.

Always check the markings on a cylinder to ascertain the identity of the gas, and ensure that you are familiar with its properties and associated hazards. Cylinders are labelled according to safety legislation, usually on the collar or shoulder and are also painted in a ground colour. Colour-code charts to aid identification are available from the gas supply companies. Additional warning bands at the valve end give an indication of hazardous flammable or toxic properties.

RED band: flammable,
YELLOW band: toxic.

A threaded outlet at the top of the cylinder incorporates the main on/ off valve, and since this serves only to turn the gas supply on or off, a compressed gas cylinder must be fitted with some means of controlling the gas pressure and/or flow rate. Diaphragm regulators (Figure 6.7) are used to provide pressures which are compatible with whatever equipment is connected to the supply, e.g. inert gas lines require about 2–3 lb inch^{-2} nitrogen or argon, while high pressure autoclaves may require several tens of atmospheres of a reagent gas (a compressor must be used when pressures of several hundreds of atmospheres are necessary). Most general purpose regulators have output pressure ranges of about 0–50 or 0–100 lb inch^{-2} but designs are available which will deliver lower (e.g. 0–25 lb inch^{-2}) or higher (e.g. 0–2500 lb inch^{-2}) pressures. A single stage of regulation is incorporated in the less expensive regulators, and the output pressure in these designs may slowly decrease as the cylinder

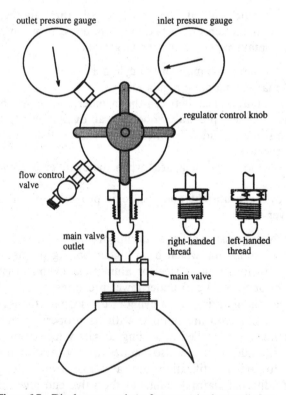

Figure 6.7 Diaphragm regulator for pressurized gas cylinders.

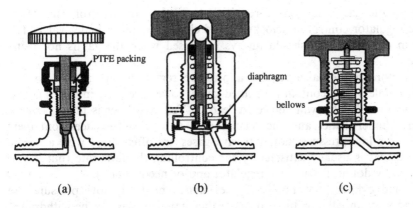

Figure 6.8 Examples of (a) needle, (b) diaphragm and (c) bellows valves.

pressure decreases. When a constant delivery pressure is important, use a two-stage regulator. To control the flow rate, a packed needle valve (Figure 6.8a) is generally connected to the low pressure side of the regulator. For high purity gases, the possibility of contamination due to leakage past the valve stem can be eliminated by using a diaphragm or bellows valve (Figures 6.8b and 6.8c, respectively). The correct procedure for fitting and using pressure regulation and flow control equipment is described below.

- Remove the valve protection cap, if fitted.
- Remove the protective cap on the threaded outlet, if fitted.
- Ensure that the threads are completely free of dirt, dust or plastic (**do not** do this by opening the main valve to blow out the solids).
- Ensure that the rounded sealing surface of the regulator is clean and screw it into the threaded outlet on the cylinder. Tighten it firmly (grease or Teflon tape should never be used on these fittings) and check for leaks with a dilute soap solution. Do not overtighten with undue force.
- Screw out the adjustment knob on the diaphragm (anticlockwise) until it rotates freely. The regulator is then closed and no pressure will appear at the outlet when the inlet is pressurized.
- Using the correct key, carefully open the main cylinder valve until the inlet gauge of the regulator registers the cylinder pressure.
- Close the flow control valve.
- Turn the regulator adjustment screw clockwise (it should get tighter as you turn it) until the required pressure is shown on the outlet pressure gauge.
- Check the connections to the apparatus and open the flow control valve slowly.

When the gas is no longer required, close the flow control valve, turn the regulator control anticlockwise and close the main cylinder valve. The main cylinder valve should always be closed when the gas is not being used.

Lecture bottles also require a pressure regulator and a flow control valve unless they contain a liquified gas at lower pressure, in which case a flow control valve will suffice. A lead or Teflon washer is fitted between the cylinder outlet and the regulator or valve. Welded steel containers used for low pressure, non-corrosive gases are often fitted with a needle valve that is usually satisfactory for controlling the flow rate (but if you have any doubt, fit an extra regulator and/or needle valve).

Some cylinders of liquified gases under relatively low pressure are fitted with an eductor tube so that either liquid or gas can be withdrawn. The liquid withdrawal port is marked with an 'L' stamped on the valve spindle or handwheel. To obtain liquid, the cylinder is laid on its side so that the end of the eductor tube is below the surface of the liquid (Figure 6.9). Smaller cylinders are often kept in a rocking cradle which can be constructed from the type of angle-iron used for shelving. The correct pressure and flow control equipment for any particular gas **must** be used, and this will be determined by the gas supply pressure, the physical and chemical properties of the gas and its purity. In particular:

- **Screw fittings** for **flammable** gases (e.g. H_2, propane, CO) have **left-handed threads** and are identified by V-shaped indentations on the edges of the tightening nuts (Figure 6.7).
- Non-corrosive fittings made from stainless steel or Monel should be used for reactive gases such as hydrogen halides and metalloid halides. It is also good practice to purge with inert gas before and after use to minimize chances of corrosion.

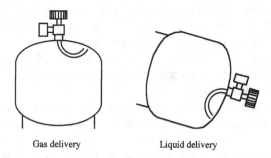

Gas delivery   Liquid delivery

Figure 6.9 Use of the eductor tube fitted to cylinders of liquid ammonia.

- Fittings for oxidizing gases such as oxygen must not contain oil, grease or other flammable materials.
- Fittings for flammable gases should be earthed (grounded) to minimize the risk of static discharge.
- Fittings for hygroscopic corrosive gases (e.g. HCl) should include provision for purging the gas line, and feedback of corrosive liquid into the cylinder should be prevented.
- Alloy fittings for acetylene must not contain more than 70% copper or 43% silver because of the risk of forming explosive acetylides. (Mercury also forms an explosive acetylide.)
- Copper and copper alloys are attacked rapidly by ethylamine, methylamine, dimethylamine and trimethylamine.

6.5.2 Generation of reagent gases

When the amount of a reagent gas required is insufficient to justify the expense of a cylinder, it may be possible to generate it in a chemical reaction and, if necessary, purify it before admitting it to the reaction flask as described in section 7.2.3 (Figure 7.5). The addition of a liquid reagent to another reactant requires fairly simple apparatus, although the incorporation of a liquid inlet valve (Figure 6.10) designed by H.C.

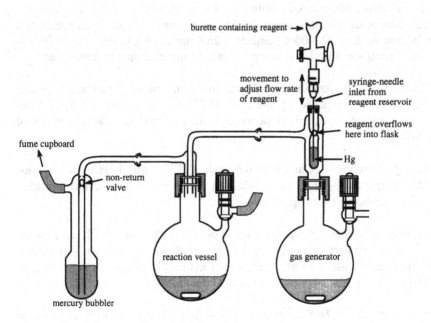

Figure 6.10 Use of a gas generator.

Brown [23,24] enables the addition to be controlled automatically. The liquid is added through a needle immersed in mercury and flows out of small holes above the mercury level, dropping onto the reactant and generating gas only when the pressure in the reaction flask drops as the gas is consumed. This apparatus is suitable for gases which do not react with mercury (e.g. H_2, CO, O_2, SO_2, hydrogen halides and ethene).

Methods for generating a selection of gases are given below and the following section describes purification procedures [1,6]. For the quantitative manipulation and measurement of gases see sections 5.3 and 7.2.3.

Acetylene. **Flammable.** Acetylene may explode spontaneously at high pressures, so when generated by the addition of water to calcium carbide, the gas generator should be kept cool and the flow of gas must not be restricted. Avoid pressures above 15 lb inch^{-2}. Do not discharge into an area where there is a risk of forming an explosive mixture with air. Waste gases should be flared through a suitable burner fitted with a flashback arrester. Forms explosive acetylides with copper, silver and mercury (do not use mercury bubblers).

Carbon dioxide. Gaseous carbon dioxide is readily prepared by evaporation of the solid or by the addition of dilute hydrochloric acid to calcium carbonate.

Carbon monoxide. **Toxic, flammable** (TLV 50 ppm). Although there are problems with frothing, slow addition of formic acid to concentrated sulphuric acid at 90–100°C generates 26.6 mmol of CO per cm^3 of formic acid. Excess gas should be vented slowly in an efficient fume cupboard.

Chlorine. **Toxic, corrosive, oxidant, damaging to the respiratory system** (TLV 1 ppm). Prepared by the slow addition of concentrated hydrochloric acid to solid potassium permanganate (6.2 cm^3 of acid per gram of $KMnO_4$).

Hydrogen bromide. **Corrosive and toxic** (TLV 3 ppm). HBr is conveniently prepared by the slow addition of bromine to stirred, purified tetralin. Dissolve excess gas in water.

Hydrogen chloride. **Corrosive and toxic** (TLV 5 ppm). HCl is prepared from concentrated sulphuric acid and anhydrous sodium or ammonium chloride or by addition of concentrated hydrochloric acid to concentrated sulphuric acid. Dissolve excess gas in water.

Sulphur dioxide. **Toxic, corrosive** (TLV 2 ppm). Prepared by addition of dilute hydrochloric acid to sodium sulphite. Dissolve excess gas in water.

6.5.3   Purification of reagent gases

Impurities should be removed from gases prepared in the laboratory, and even when used from a cylinder, some gases will need to be purified.

Acetylene. Acetone is added to acetylene cylinders to reduce the pressure, and this can be removed by passing the gas through an aqueous solution of $NaHSO_3$ and then drying it with molecular sieves. Acetylene prepared from calcium carbide can similarly be dried by passing it through a column of molecular sieves.

Ammonia. Dry by passing the gas through a solution of aluminium isopropoxide in toluene or a higher boiling hydrocarbon solvent. Any entrained hydrocarbon will usually not cause problems in subsequent reactions, but should its presence be undesirable, the ammonia can instead be dried on a vacuum line by condensation onto sodium and then distillation into the reaction flask.

Carbon dioxide. The usual contaminant in carbon dioxide is water and this can be removed on the vacuum line by trap-to-trap sublimation through an activated silica gel trap.

Carbon monoxide. Oxygen can be removed from carbon monoxide by passing the gas through a column of supported copper (of the type used for nitrogen purification in glove boxes) or supported MnO. It has been reported that the MnO is not readily regenerated once exposed to CO [6].

Ethene. Very pure, oxygen-free ethene (and other volatile alkenes) for reactivity studies can be obtained by passing the gas through supported MnO [25].

Hydrogen. Very pure hydrogen is obtained by diffusion through heated palladium and because of its use as a carrier gas in the electronics industry, commercial units are readily available (although somewhat expensive).

Hydrogen halides. Trap-to-trap distillation of HCl and HBr at 100 mmHg or below enables water to be condensed in a trap held at $-78°C$. Halogen impurities can be adsorbed onto an activated carbon trap, while gases that are not condensable at liquid nitrogen temperature can be removed by transfer under dynamic vacuum.

Sulphur dioxide. Atmospheric gases can be removed from the commercial compressed gas by condensation at liquid nitrogen temperatures and pumping on the vacuum line. Water is removed by passing the gas through a trap containing P_4O_{10} supported on glass wool.

References

1. Casey, M., Leonard, J., Lygo, B. and Procter, G. (1990) *Advanced Practical Organic Chemistry*, Blackie, Glasgow.
2. Loewenthal, H.J.E. and Zass, E. (1990) *A Guide for the Perplexed Organic Experimentalist*, 2nd edn, John Wiley & Sons, Chichester.
3. Keese, R., Müller, R.K. and Toube, T.P. (1982) *Fundamentals of Preparative Organic Chemistry*, Ellis Horwood Ltd, Chichester.
4. Perrin, D.D. and Armarego, W.L.F. (1988) *Purification of Laboratory Chemicals*, 3rd edn, Pergamon Press, Oxford.
5. Riddick, J.A., Bunger, W.B. and Sakano, T.K. (1986) *Organic Solvents*, 4th edn, John Wiley and Sons, New York.
6. Shriver, D.F. and Drezdzon, M.A. (1986) *The Manipulation of Air-Sensitive Compounds*, 2nd edn, Wiley, New York.
7. Wayda, A.L. and Darensbourg, M.Y. (eds) (1987) *Experimental Organometallic Chemistry: a Practicum in Synthesis and Characterization*, ACS Symposium Series, Vol. 357.
8. Salmon, J., Chapman, R.F. and Lammin, S. (1995) *Chem. Br.*, **31**, 942.
9. Burger, B.J. and Bercaw, J.E. (1987) In *Experimental Organometallic Chemistry: a Practicum in Synthesis and Characterization*, eds A.L. Wayda and M.Y. Darensbourg, ACS Symposium Series, Vol. 357, p.79.
10. Pez, G.P. and Armor, J.N. (1981) *Adv. Organomet. Chem.*, **19**, 1.
11. Burfield, D.R. and Smithers, R.H. (1978) *J. Org. Chem.*, **43**, 3966.
12. Wayda, A.L. and Bianconi, P.A. (1987) In *Experimental Organometallic Chemistry: a Practicum in Synthesis and Characterization*, eds A.L. Wayda and M.Y. Darensbourg, ACS Symposium Series, Vol. 357, p. 76.
13 Felkin, H. (1987) In *Experimental Organometallic Chemistry: a Practicum in Synthesis and Characterization*, eds A.L. Wayda and M.Y. Darensbourg, ACS Symposium Series, Vol. 357, p. 74.
14. *Inorg. Synth.*, (1986) **24**, pp 1–81.
15. Parkington, M.J., Seddon, K.R. and Ryan, T.A. (1989) *J. Chem. Soc., Chem. Commun.*, 1823.
16. Suffert, J. (1989) *J. Org. Chem.*, **54**, 509.
17. Watson, S.C. and Eastman, J.F. (1967) *J. Organomet. Chem.*, **9**, 165.
18. Connelly, N.G. and Geiger, W.E. (1996) *Chem. Rev.*, **96**, 877.
19. Fürstner, A. (1993) *Angew. Chem. Int. Ed. Eng.*, **32**, 164.
20. Araya, M.A., Cotton, F.A., Matonic, J.H. and Murillo, C.A, (1995) *Inorg. Chem.*, **34**, 5424.
21. Kilner, M. and Parkin, G. (1986) *J. Organomet. Chem.*, **302**, 181; Kilner, M., Parkin, G. and Talbot, A.G. (1985) *J. Chem. Soc., Chem. Commun.*, 34.
22. Coulson, D.R. (1990) *Inorg. Synth.*, **28**, 107.
23. Brown, C.A. and Brown, H.C. (1966) *J. Org. Chem.*, **31**, 3989.
24. Shriver, D.F. and Drezdzon, M.A. (1986) *The Manipulation of Air-Sensitive Compounds*, 2nd edn, Wiley, New York, p. 25.
25. He, M.-Y., Xiong, G., Toscano, P.J., Burwell, Jnr., R.L. and Marks, T.J. (1985) *J. Am. Chem. Soc.*, **107**, 641.

7 Carrying Out Reactions in Solution

7.1 Introduction and general considerations

Most reactions are carried out with at least one of the reactants in solution because this is the easiest way to achieve intimate mixing and temperature control. This section, therefore, takes you through the various stages that will be involved in the vast majority of your reactions, especially those involving air-sensitive compounds. From writing down an equation for the reaction to product characterization the routines will eventually become second nature, but it is important that you develop a thorough approach from the outset and try to eliminate any bad habits.

Whether it is a new reaction or a repeat of a literature procedure, having written down an equation and determined the amounts of each reagent to be used, you **must** first consider the safety aspects. The properties of the reagents, solvents and products, the scale and likely nature of the reaction (will it be exothermic or evolve a gas?), the necessary precautions, control measures and disposal procedures should all be documented before you start. In the UK, remember that this is a legal requirement under the COSHH regulations (section 2.4).

New reactions should be tried on a small scale first to give you some idea of how they proceed. Don't get overzealous when scaling up a reaction; exothermic processes can pose problems on a large scale because of heat transfer limitations. An induction period followed by a rapid increase in temperature may result in a reaction that you cannot control.

Once you are familiar with the compounds involved, try to visualize yourself carrying out each step of the reaction. Plan ahead, anticipate any problems and ensure that all the necessary apparatus (including stirrer bars, stoppers, syringes, cannulae, etc.) is available and that it is clean and dry. In particular, you need to make several decisions that may be crucial to the success of the reaction.

- What order to add the reactants
- How to add the reactants
- What solvent to use
- Whether to cool/heat the reaction mixture
- How to isolate the products

The following sections in this chapter should help you address the first four of these questions, while product isolation is dealt with in chapter 8.

Proper attention to the above preliminaries will ensure that you are prepared for most eventualities before you begin a reaction, and should also help you develop the confidence that is necessary if your time in the laboratory is to be fruitful.

7.2 Measuring out reagents

Given the fundamental importance of having the correct stoichiometry in a reaction, it is somewhat surprising that many students make their first mistake before they even reach for the chemicals. Check all your calculations of formula weights and amounts **before** measuring out reagents. A mistake here could waste a lot of time later!

7.2.1 Solids

It is best if you can add the solid reactants to the reaction flask first, since the manipulations involved become awkward once there is liquid in the flask. If a solid reactant must be added later during the reaction, it can be weighed into a separate flask, dissolved in an appropriate solvent and then added to the main reaction flask as a solution *via* cannula (section 3.4.1). Alternatively, a solids addition tube (Figure 3.16a) can be used as shown in Figure 7.1. As the tube is rotated upwards, the solid drops into the flask.

Weighing a solid into the reaction flask or into an addition tube presents no problems if the compound is air-stable and relatively non-toxic, although the size or shape of larger pieces of glassware can make it awkward to position them on the weighing pan of a bench-top balance. Therefore, it is often easiest to first weigh the solid into a weighing boat or a sample tube or onto piece of paper (preferably non-absorbent) and then transfer it to the reaction flask or addition tube. However, air-sensitive or toxic solids need to be contained while they are being weighed, and this is easiest if you have access to a balance in a dry box. The compound is taken into the box with the necessary glassware and a weighing boat (or aluminium weighing pan/foil) and accurate amounts can then be weighed out and transferred into the glassware. If such a facility is not available, the solid must be weighed by difference as follows.

- Attach a clean, dry flask to the line and evacuate/flush with nitrogen a few times.
- Close the tap and record the weight of the flask (either evacuated or filled with nitrogen, but make sure you make a note of which).
- Either in a dry box or by attaching the storage flask at your line, add

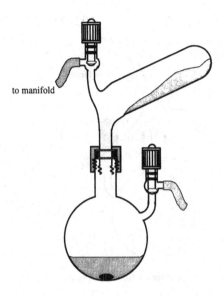

to manifold

Figure 7.1 Addition of solids to a reaction.

solid to the flask under nitrogen, estimating the amount necessary to give you the required mass.

• Make sure that the flask is sealed and the tap is closed and record the weight of the flask (either evacuated or filled with nitrogen as above) plus the solid.

• The weight of solid is the difference between the two recorded weights.

When using this method it is best to calculate the required amounts of all other reagents from this weighed quantity of solid; otherwise you will have to go through the whole process several times to adjust the amount in the flask to some pre-determined value. However, when weighing out two (or more) solids for a reaction by difference, you will almost certainly have to go through the adjustment process for one (or more) of the compounds unless you are very good at guessing how much you have added.

If your starting material is in a sealed ampoule and you intend to use it all in the reaction, it is possible to open the ampoule and add the contents to the reaction flask at your bench rather than in the dry box, provided the compound is not too air-sensitive.

• Connect a filter funnel to the nitrogen line and arrange it over the neck of the flask as shown in Figure 7.2.

• Using a glass-cutting knife, score the ampoule and if the weight of the contents is not known, weigh it.

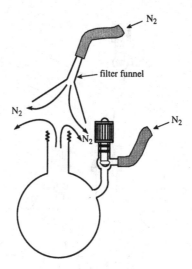

Figure 7.2   Use of an inverted funnel to provide an inert-gas 'blanket'.

- Open the taps on the line and the flask to allow inert gas to flow from the neck of the flask and from the funnel.
- Remove the stopper on the flask and turn up the nitrogen.
- Open the ampoule in the combined gas streams emanating from the flask and the funnel and then quickly insert it into the neck of the flask. The tap on the line connected to the funnel can then be closed and the nitrogen flow reduced.
- Empty the contents into the flask with gentle tapping.
- If necessary, re-weigh the two parts of the empty ampoule and calculate how much solid has been added.

7.2.2   Liquids

Liquids are most easily handled by syringe and cannula techniques (section 3.4) and measured by volume, so you need to know the density of a pure compound or the concentration of a solution. If the density is not available, you can measure it using the procedure outlined below.

- Fill a syringe to a graduation mark and seal the needle by sticking it into a solid rubber bung.
- Record the weight of the whole assembly.
- Eject a known volume from the syringe but leave the needle filled with liquid, re-sealing it with the solid rubber bung.
- Record the new weight of the assembly.

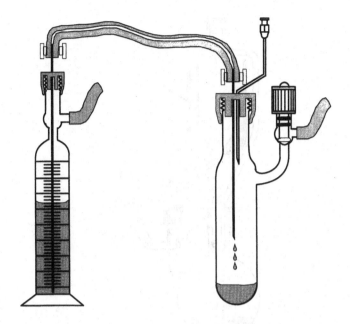

Figure 7.3   Transferring a measured amount of liquid from a calibrated vessel.

- The difference between the two values is the weight of the known volume of liquid, enabling the density to be calculated.

Volumes of up to 50 cm^3 can be handled with a syringe, but larger volumes are best added from a calibrated container *via* cannula (Figure 7.3). It is sometimes inconvenient to add a very small volume from a microlitre syringe and in such cases it is worth making up a dilute solution of known concentration so that a more convenient volume can be added from a larger syringe.

When a liquid must be added over an extended period of time a dropping funnel is usually used, and one with cooling jacket allows low temperature addition (Figure 7.4). Measured volumes are transferred to the dropping funnel from a storage vessel or calibrated flask by cannula or syringe. For slow addition of smaller volumes from a syringe, electric motor drives which push the piston slowly into the barrel at a controlled rate are available commercially.

7.2.3   Gases

If an excess of a gas is to be added from a cylinder or generator (section 6.5), it can be led directly into the reaction mixture through a delivery

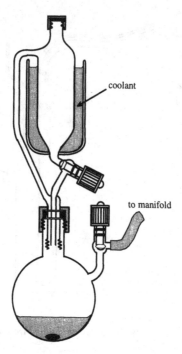

Figure 7.4 Use of a pressure-compensating dropping funnel fitted with a cooling jacket.

tube as shown in Figure 7.5. This type of reaction should be carried out in a fume cupboard and it is important to include an oil or mercury bubbler before the reaction flask to prevent a dangerous build-up of pressure should the delivery tube become blocked. An empty trap protects against suck-back, especially for more soluble gases, and unreacted gas exits the reaction flask through a bubbler into the fume cupboard. If the gas is toxic the bubbler should be connected to a scrubber to prevent release of the gas into the atmosphere. Connecting tubing should be compatible with the type of gas being used and while rubber or PVC can be used for non-toxic, non-corrosive gases, other materials such as Teflon or Viton may be necessary for reactive gases.

For quantitative work, volatile liquids and condensable gases can be measured into the reaction flask using a calibrated vacuum manifold as described in section 5.3, but a simpler method of transferring a measured volume of gas into a reaction flask is to use a gas burette as described below and shown in Figure 7.6.

- Turn the three-way tap A to connect the burette to the supply line.
- Open tap B to the fume cupboard and raise the reservoir to expel all the gas from the burette.

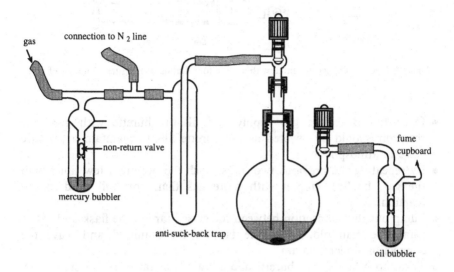

Figure 7.5   Addition of a gas to a reaction.

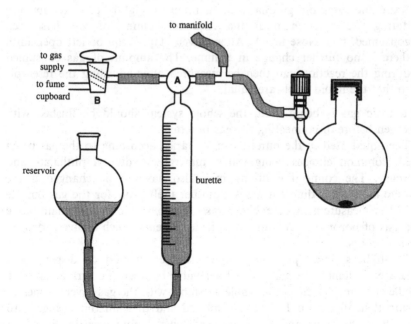

Figure 7.6   Addition of a gas from a gas burette.

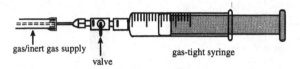

gas/inert gas supply

valve

gas-tight syringe

Figure 7.7 Gas-tight syringe fitted with a valve for addition of measured volumes of gas.

- Open tap B to the gas supply and fill the burette with gas. The reservoir should be lowered as the burette fills to maintain a pressure close to atmospheric.
- By repeating the previous two steps, flush the burette a few times with the gas, finally filling it with more gas than you will need in the reaction.
- Pump out the connection between the closed tap on the flask and tap A using the manifold. Close the tap on the manifold and leave the connection under vacuum.
- Turn tap A to seal the burette and allow gas as far as the closed tap on the flask.
- Turn off tap B.
- Turn tap A to connect the burette to the flask, adjust the reservoir slightly to ensure that the gas in the burette is at atmospheric pressure. Note the reading on the burette and open the tap to the reaction flask.
- Raise the reservoir gradually to maintain a slightly positive pressure during the reaction until the required volume of gas has been consumed, then close tap A. Alternatively, tap A can be left open until there is no further change in volume. The amount of gas consumed during the reaction can then be measured after adjusting the reservoir so that the liquid levels are equal.

If a toxic gas is being used, the whole system should be flushed with nitrogen before disconnecting the gas burette.

The liquid used in the burette can be varied according to the gas being used, common choices being water, mineral oil, dibutyl phthalate and mercury. The connecting tubing may also have to be changed if the burette is used for different gases. Another useful role for the gas burette is in the measurement of evolved gases, for example in the quantitative analysis of boron and aluminium hydride reagents which evolve hydrogen on hydrolysis.

Gas-tight syringes provide an even simpler method of dispensing a measured volume of a gas. Fit a Luer valve between the syringe and the needle (Figure 7.7), flush the whole assembly with the gas several times to ensure that all the air has been displaced and then fill the syringe with gas. Open the valve and flush the needle just before inserting it into a septum cap on the reaction flask.

7.3 Transferring solvents

Your choice of a solvent for a reaction will normally be based upon your past experience or on information from related literature procedures, but in the absence of both there are some general guidelines that can be followed. The direct involvement of the solvent in the reaction is usually avoided, except when it may act as a ligand to a metal centre, and even this is undesirable in some cases. The choice may also be made on the basis of polarity; hydrocarbons will usually dissolve non-polar complexes such as metal alkyls whereas ionic complexes will require more polar solvents such as acetone or acetonitrile. The more volatile solvents will be easier to remove during work-up, but if elevated temperatures are required during the reaction a higher boiling solvent may be necessary. Selected solvent properties are given in Appendix C.

If the chosen solvent for a reaction is one of those in the permanent stills, then the requisite amount can be transferred by syringe to the reaction flask, although for larger volumes this may entail several trips between your bench and the still. To eliminate multiple transfers, some still designs allow the solvent to be run directly into the flask under nitrogen (section 6.3). Other solvents will have to be dried and degassed on a batch basis, stored under nitrogen and then transferred from a storage flask to the reaction flask under nitrogen by syringe or cannulation.

When adding the solvent, try to direct the syringe needle or cannula around the walls of the flask in a circular fashion to wash any solids to the bottom. Aim to keep the walls of the flask above the liquid level clean throughout the reaction so you can be confident that all of the starting materials are in solution and have a chance to react. The total volume of solvent should not exceed 40–50% of the volume of the flask. This will make it easier to control the reaction, especially if it is exothermic and/or a gas is evolved. Reaction work-up will also be easier and bumping will be less of a problem during the removal of solvent under reduced pressure.

7.4 Mixing

If the reaction mixture is a homogeneous solution (i.e. all reactants dissolve and the products are soluble) then stirring is not strictly necessary unless heating is involved, when it will help to prevent bumping. However, it is not always possible to predict the course of a reaction, and most reactions have solid present at some stage, so it is best to provide a means of stirring to ensure thorough mixing.

7.4.1 Magnetic stirring

Magnetic stirrers and stirrer/hotplates are readily available and are the method of choice for small- to medium-scale reactions unless the mixture is very viscous. In these stirrers, a permanent magnet mounted below a flat, non-magnetic surface (which may or may not be heated) is driven by an electric motor. Depending on the design, stirrer speed control is achieved either mechanically or electronically, although the reliability of different models can vary enormously.

Magnetic followers (stirrer bars) placed in the reaction flask must be inert to the various components of the reaction mixture and are usually coated in Teflon. They are available in a wide range of sizes and shapes (Figure 7.8) and it is important to use the optimum size for the reaction flask. An oversize follower will not stir in the bottom of the reaction flask, tending instead to jump around especially when the stirring speed is increased. In this situation, there is a serious risk that the follower will break the walls of the flask and this risk is greater with larger stirrer bars. Egg-shaped followers designed for round-bottomed vessels work well for most reactions in Schlenk flasks, whereas octagonal followers tend to be more effective in flat-bottomed beakers and conical flasks. The smaller magnetic 'fleas' tend to be plain oval bars. Glass-encapsulated stirrer bars are useful for reactions involving strong reducing agents, e.g. sodium naphthalenide, which attack Teflon. These can be made by sealing a length of small diameter iron or steel rod (this can be cut from a nail) or the innards of a Teflon-covered follower in glass tubing.

If the reaction flask must be placed inside a heating mantle or a Dewar flask, a magnetic stirrer is likely to be ineffective because of the distance between the drive magnet and the follower. In these situations a stirrer-heating mantle or a remote, submersible magnetic stirrer (Trömner stirrer) provide the ideal solutions.

7.4.2 Mechanical stirring

For viscous mixtures and large-scale reactions, a mechanical, motor-driven stirrer must be used. Note that sparks from electric motors present a fire hazard in the presence of flammable solvents, although

plain bar   egg-shaped   octagonal

Figure 7.8 Magnetic stir-bars.

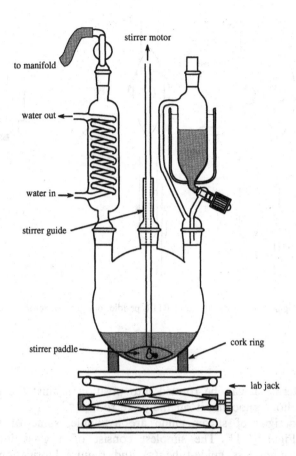

Figure 7.9 Typical reaction setup involving liquid addition and mechanical stirring.

spark-free motors are available or a mechanical stirrer driven by compressed air can be used. The motor is mounted above the reaction flask and is connected to a paddle inside the flask by a rigid (glass) or flexible (Teflon) rod. In the most common design, a detachable crescent-shaped Teflon paddle is connected to the end of the rod by a shaped peg that prevents the paddle falling off once it has been rotated into position. Figure 7.9 shows a typical setup for a large-scale preparation involving mechanical stirring, dropwise addition of a reagent and reflux. Note the use of a lab-jack which enables heating/cooling baths to be raised and lowered without disturbing the glassware and stirrer. Details of the paddle attachment are shown in Figure 7.10. The paddle should be small enough to pass through the neck of the flask after which, with

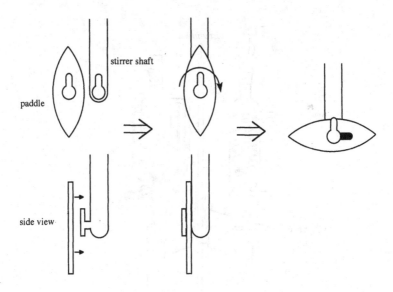

Figure 7.10   Attachment of PTFE paddle to a glass stirrer shaft.

a little dexterity, it can be rotated into position against the bottom of the flask without knocking it off the drive rod.

Various designs of stirrer guide are available, some of which are shown in Figure 7.11. The simplest consist of a glass joint and a precision ground-glass guide-tube (a) and require lubrication, but for inert atmosphere work an efficient seal that allows rotation without leakage is required. Older glass designs use a mercury seal (b) but a build-up of pressure in the flask can cause vapours to bubble out through the mercury (especially if the height of mercury in the bubbler on the nitrogen manifold is higher than that in the stirrer guide). Better modern versions (c) are constructed using a mixture of Teflon and other plastics and employ O-ring seals (and, of course, they are more expensive).

When setting up apparatus similar to that shown in Figure 7.9, it is important to take the time to align all the components properly so as to minimize stresses and vibrations. Clamps should hold the various parts in compression and should be double-checked before beginning the reaction. Before adding any chemicals also ensure that the stirrer paddle rotates freely at the correct height and that the lab-jack can be adjusted to allow heating/cooling equipment to be added or removed easily.

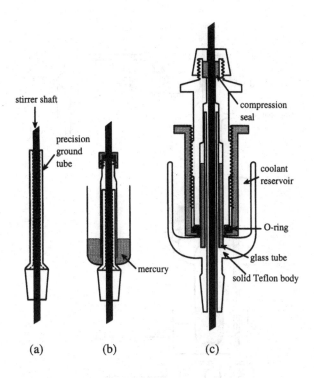

stirrer shaft

precision
ground
tube

compression
seal

coolant
reservoir

O-ring

mercury

glass tube
solid Teflon body

(a)   (b)   (c)

Figure 7.11   Stirrer guides and seals.

7.5   Heating

Reaction mixtures are heated in order to increase the rate of reaction or
to dissolve a compound that is not soluble at ambient temperature. When
heating is necessary, you need first to consider the maximum temperature
required and the likely duration of the heating period (these factors
should already have been considered when choosing a solvent and the
type of apparatus to be used). Some heating options are described below.

7.5.1   Heating under reflux

For extended heating periods it is usual to choose a solvent having a
boiling point sufficiently high to provide a satisfactory reaction rate
without causing decomposition of reactants or products. The reaction can
then be heated under reflux conditions for as long as is necessary. A
simple reflux apparatus is shown in Figure 7.12a and a typical procedure
for setting it up under nitrogen is given below.

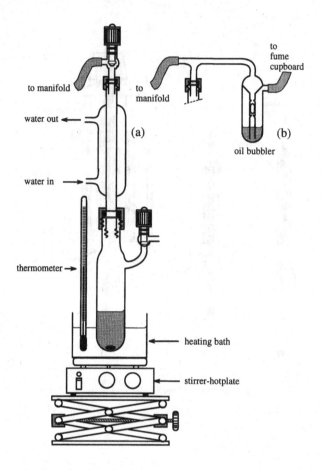

Figure 7.12 Typical setup for heating a reaction mixture under reflux.

- Add reactants and solvent to the reaction flask as described above.
- Attach a tap and end-cap to a clean, oven-dried condenser and pump it under vacuum.
- Flush the condenser several times with nitrogen.
- Establish nitrogen flows through the condenser and the flask while removing their caps/stoppers, attach the condenser to the flask and ensure that there are no leaks (check the bubbler on the nitrogen line).
- Close the tap on the flask.
- Turn on the cooling water (if necessary).
- Heat the flask gradually until the solvent refluxes.

Condensers. The simple Liebig condenser shown in Figure 7.12a will be sufficient for most small- to medium-scale reactions, but for larger scale reactions involving volatile solvents or gas evolution a more efficient condenser may be necessary. Typical water-cooled coil and double-surface designs are shown in Figure 7.13 (see also Figure 7.9). It is also well worth considering the use of air-cooled Liebig condensers, which use metal heat exchangers instead of water cooling, to eliminate any risk of flooding during extended reflux. In reactions where a gas is evolved, it can be swept out of the apparatus by attaching a bubbler to the top of the condenser and maintaining a slow stream of nitrogen (Figure 7.12b) through the system. The exhaust should be led to the fume cupboard or to a scrubber.

When using a Schlenk tube, it is possible to heat higher boiling solvents under gentle reflux without fitting a condenser, the vapours will condense on the walls of the flask provided the flask is somewhat less than one quarter full. Note that the flask should be open to the line.

Heating baths. Controlled heating is best achieved with a heating bath on an electric hotplate. The material used in the bath will depend on the temperature required, but the most common are water, mineral oil, silicone oil, flake graphite and sand. Water baths are obviously limited to temperatures up to 100°C and are not the best option when working with moisture-sensitive compounds. Mineral oil can be taken up to about 180°C, but starts to decompose and smoke at higher temperatures and gradually darkens on repeated use. It should be discarded (into an

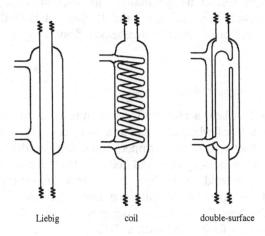

Liebig   coil   double-surface

Figure 7.13 Water-cooled condensers.

oil residue container) when it becomes very murky. Although compara-tively expensive, silicone oils are much more stable and are available in several fractions with temperature ranges up to about 250°C. After removing a flask from an oil bath, let the oil drain back into the bath and then carefully wipe off any remaining oil, otherwise things can get messy and you will find oil smeared over the rest of your glassware. Graphite baths are convenient, can be used up to 300°C and, in theory, are less messy than oil baths, but in practice the graphite flakes are readily blown out of the bath by draughts and also tend to adhere to the walls of the glassware. Sand can be used instead, but requires a high heat input from the hotplate and takes a longer time to cool. A selection of other materials can be used for higher temperatures. Wood's metal, an alloy (50% Bi, 25% Pb, 12.5% Sn and 12.5% Cd) that melts at 70°C, is convenient for temperatures up to 300°C although concerns about toxicity mean that it should be used in the fume cupboard. Dibutyl phthalate has a boiling point of 340°C, and salt mixtures, e.g. 40% $NaNO_2$, 7% $NaNO_3$, 53% KNO_3 (m.p. 142°C) and 51.3% KNO_3, 48.7% $NaNO_3$ (m.p. 219°C), can be used up to 500°C.

When using a heating bath, bear in mind that the temperature of the contents of the flask will be lower than that of the bath, and bath temperatures of about 30°C higher than the boiling point of the solvent will generally be required to bring the reaction mixture to reflux.

Heating mantles. Flasks can be heated directly using a heating mantle but the correct size (and shape!) mantle must be used (they are normally designed only for round-bottomed flasks) to prevent damage to the heating element through overheating. Temperature control is more difficult than with a heating bath and hot-spots may develop, so the heat control should be turned up gradually. The level of liquid in the flask should not be lower than the top of the heating mantle; otherwise some decomposition may occur on the overheated flask wall above the liquid level.

7.5.2 Heating for short times

If you only want to heat a mixture for a minute or so to dissolve a solid or to get some idea of whether a reaction will occur upon increasing the temperature, a heat gun (paint-stripper, hair-dryer) is ideal. In this case you can control the heating to avoid reflux so that the solvent does not enter the hose connected to the line. If the flask has ground-glass joints the flask must be open to the nitrogen line otherwise the increase in pressure will force out the stopper allowing air to enter the flask, whereas with a screw-capped flask fitted with a Teflon screw tap, the flask can be sealed and heated up to and slightly beyond the boiling point of the solvent without creating a dangerous pressure in the flask. However, **exercise great care** when heating a sealed system like this. Make sure that there is no likelihood of a gas being evolved on heating and use a safety screen. Further details of reactions in sealed tubes and pressure vessels are given in chapter 12, as are details of microwave heating, and the use of photochemical and ultrasonic activation. Infrared heating is discussed in sections 6.2 and 8.3.

7.6 Cooling

Exothermic reactions are usually controlled by cooling the reaction mixture to below room temperature in a cooling bath. You should immerse the flask far enough into the bath to ensure that all the contents are beneath the level of the coolant and use a low-temperature thermometer or thermocouple probe to monitor the temperature. It is easier to sustain a low temperature if a Dewar flask or some other form of insulated container is used for the bath and, bearing in mind that you will probably want to clamp the flask and stir the reaction mixture while maintaining easy access for the addition of reagents, a low 'squat-form' shape that will fit on a magnetic stirrer is best. Commercially available Dewars of this type tend to be expensive, although our glassblowers make them from round-bottomed flasks (1 l or smaller) which, after silvering, are mounted in aluminium bases. The construction of an even cheaper alternative (Figure 7.14) from two nested dishes of glass, polypropylene or polycarbonate with polyurethane foam insulation (finely divided vermiculite packing material could also be used) has been described [1].

With the dry-ice (solid CO_2) and liquid nitrogen slush baths described in section 5.3 and listed in Table 5.1 temperatures down to $-196°C$ can be obtained. Sub-zero temperatures (from $0°C$ to $-40°C$) can also be obtained by adding various salts or solvents to ice (Table 7.1) although at the lower end of this range very little liquid is present resulting in inefficient heat transfer, so slush baths are better for these temperatures.

It is worth remembering that quoted temperatures for cooling baths are the minimum obtainable and, as with heating baths, there will be a difference between the temperature of the bath and that of the contents of the flask, so when careful temperature control is crucial the temperature of the reaction mixture should be monitored.

Although cooling baths can maintain low temperatures for several hours if properly insulated and lagged, longer term temperature stability is best achieved with a commercial refrigeration unit. These portable systems have a cooling probe/coil that can be immersed in a

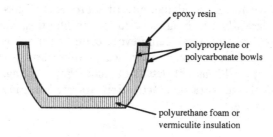

Figure 7.14   An insulated cooling bath.

Table 7.1   Ice-based cooling baths[a]

Additive	Ratio (ice/additive)	Temperature/°C
Water	1:1	0
NaCl	3:1	−8
Acetone	1:1	−10
$CaCl_2 \cdot 6H_2O$	4:5	−40

[a]Gordon, A.J. and Ford, R.A. (1972) *The Chemist's Companion*, J. Wiley and Sons, New York.

suitable liquid and can give very stable, low temperatures for extended periods.

7.7   Monitoring the reaction

Once you have added all the reactants and started stirring and possibly heating the reaction mixture, how do you know when to stop? This is where careful observation pays dividends and you should record all visible changes that might indicate when a reaction has gone to completion. In addition to the qualitative colour and temperature changes or gas evolution, the progress of the reaction can be monitored more quantitatively by spectroscopic methods if the starting materials and/or products have suitable spectroscopic 'handles'. The most convenient techniques are IR, NMR and UV/visible spectroscopy. Hence, in organometallic chemistry, IR spectroscopy is ideal for studying metal carbonyl reactivity and ^{31}P-NMR is very useful for reactions involving organophosphorus ligands/reagents. Samples withdrawn from the reaction mixture are added to the appropriate cell or tube under nitrogen as described in chapter 11. When there is no further change in the spectrum, the reaction can be worked up as described in chapter 8.

It is perhaps worth mentioning here that thin layer chromatography (TLC) is not as useful in metalorganic chemistry as it is in organic chemistry, where it is routinely used to monitor the progress of reactions [1,2]. Apart from the difficulties in using this technique for air-sensitive compounds, many metal complexes are absorbed too strongly or decompose on the common supports (usually silica and alumina). Having said this, it is probably the case that most research workers don't try chromatography with metalorganic compounds because of the significant investment in time necessary to find a suitable support/eluant combination so, if your reactions are giving mixtures of closely related products that are difficult to separate by other techniques, and you have the time, it might be worth trying some TLC on the reaction mixture after stripping off the volatiles.

The majority of reactions are carried out without any spectroscopic monitoring and the reaction times you see quoted in the literature are usually convenient periods determined by the particular worker's starting time in the morning, the time and duration of any lunch or coffee breaks and the finishing time in the evening. Hence, 'overnight' could mean anything from 7 h to 17 h!

Given the above considerations and the investment in time and effort required for a reaction on a preparative scale, it is advisable, if at all possible, to first try it on a small scale with *in situ* spectroscopic monitoring (NMR and IR are the most convenient and details are given in the relevant sections of chapter 11). This can provide much valuable information about the feasibility, optimum conditions and outcome of a reaction before it is scaled up to preparative scale and, in favourable circumstances, products can be isolated and fully characterized without the need for a larger scale reaction.

References

1. Loewenthal, H.J.E. (1990) *A Guide for the Perplexed Organic Experimentalist*, 2nd edn, John Wiley and Sons, Chichester.
2. Casey, M., Leonard, J., Lygo, B. and Procter, G. (1990) *Advanced Practical Organic Chemistry*, Blackie, Glasgow and London.

8 Reaction Work-Up

8.1 Introduction

Once you have mastered the necessary techniques described in chapter 7, carrying out reactions is likely to be the easiest part of your project and, as discussed in section 7.7, the time you allow a reaction before work-up is very much a matter of your judgement. 'Work-up' is the collective term used for the procedures involved in the isolation and purification of reaction products, and this process generally requires greater initiative and takes up considerably more time than actually carrying out the reaction. If you are dealing with known compounds with documented properties or you are following an established procedure, work-up may be fairly straightforward, but when you are developing new reactions or making new compounds you must make decisions during the work-up based upon your knowledge of the physical and chemical properties of the all the compounds involved. The ability to estimate/predict thermal stability, volatility, solubility and reactivity is particularly important, and during your time in the laboratory you will find it easier to develop these skills if you talk to colleagues, learn from their expertise and ensure that you keep abreast of the relevant literature.

The true art of the synthetic chemist is revealed during the work-up stage of a reaction. Each step in the removal of residual starting materials, impurities and solvents and in the isolation and purification of individual reaction products demands attention to detail and you will be judged by your ability to apply a range of techniques to the preparation of pure compounds. Most metalorganic compounds are either solid or liquid at room temperature and separation procedures usually take advantage of differences in the solubilities and/or volatilities of the various components of the mixture.

8.2 Isolating products

It is standard practice to 'quench' organic reactions to deactivate residual reagents and facilitate subsequent product separation, but this is not usually the case in metalorganic reactions since the majority of quenching reagents tend to be aqueous solutions or contain species capable of bonding to the metal. However, ligand preparations often involve mainly

organic chemistry, and in these cases you can follow the literature procedures or the guidelines given in practical organic texts [1,2]. For example,
saturated aqueous ammonium chloride is commonly used to quench
reactions involving lithium alkyls, Grignard reagents, alkyl cuprates and
aluminium hydride reducing agents. Saturated aqueous sodium sulphate
has been suggested as an alternative quench for aluminium hydride
reagents since a denser precipitate is formed which is more easily
separated.

A solid present at the end of the reaction period can be separated by
filtration, while removal of some or all of the solvent enables the more
soluble products remaining in solution to be isolated. Alternatively,
solid products can be precipitated by addition of a solvent in which they
are sparingly soluble. These procedures are described in more detail
below.

8.2.1 Filtration

Filter-stick techniques for inert-atmosphere filtration have been described
in detail in section 3.4.2, so this section will deal with the use of a
sintered-glass filter of the type shown in Figure 3.16c. These are
constructed with medium to fine porosity glass sinters (frits), although
finely divided solids tend to block the finer pores more easily (see
comments below on the use of Celite). Remember that sintered filters
must be properly cleaned and dried before use (see section 2.1.7).

- Assemble the filter and a receiver flask while still hot from the drying
  oven, attach to the manifold and evacuate (Figure 8.1a). When cool,
  flush the apparatus with nitrogen and then repeat the purging process a
  couple of times.
- Turn up the nitrogen flow, remove the stoppers and connect the
  reaction flask to the filter so that all the taps are aligned on the same
  side (Figure 8.1b). Make sure there are no leaks and that the two flasks
  are securely connected to the filter. When the solvent is volatile, an
  increase in temperature can easily produce a pressure build-up in the
  sealed system which will cause ground-glass joints to separate unless
  they are held tightly with springs or rubber bands.
- With both taps on the filter open to nitrogen, close the tap on the
  reaction flask and remove the hose. This makes subsequent manipulation of the apparatus much easier.
- Hold the apparatus in two hands and, while swirling the reaction flask
  to suspend the solid in the liquid, carefully tilt the apparatus with the
  taps pointing upwards so that the suspension pours into the filter
  without entering the side tube.
- Gradually bring the apparatus to the vertical position and allow the

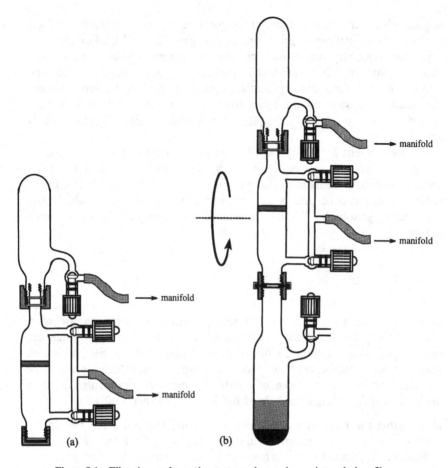

Figure 8.1   Filtration under an inert atmosphere using a sintered-glass filter.

solid to settle out under gravity (Figure 8.1c). If the solid is finely divided, it is important to keep both taps open to nitrogen at this stage, as a pressure differential may force some of the fine solid through the sinter.

- Once a pad of solid has formed on the sinter, the filtration rate can be increased by a slight reduction in the pressure below the sinter. This is best achieved by closing both taps on the filter, evacuating the connecting hose and, after closing the tap on the manifold, opening the lower tap on the filter. This procedure prevents too great a pressure reduction which may cause finely divided solid to enter the receiver or block the sinter.

- When all the liquid has passed through into the receiver, the solid on

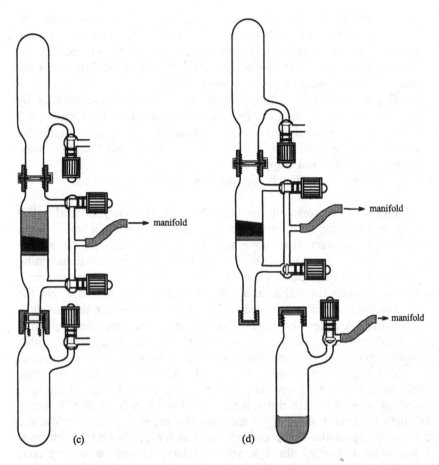

(c)   (d)

Figure 8.1 Continued.

the sinter can be washed with fresh solvent. The design shown in Figure
3.16c allows solvent to be introduced through the side tube after
removal of the screw tap, otherwise a hose must be connected to the
upper flask and the flask disconnected under a nitrogen stream to allow
solvent addition. If the solvent is volatile and you don't want to
increase the volume, freeze the filtrate in liquid nitrogen, evacuate the
system and close both taps. Wrap the empty flask and the upper part
of the filter in a cloth and **carefully** pour liquid nitrogen onto the cloth.
By gently warming the filtrate (it helps if you swirl the liquid in the
flask) the solvent will boil, condense onto the sinter and wash any
remaining solid. Cool the receiver flask slightly to return the washings
to the filtrate. Repeat this process until no more material dissolves.

- After washing the solid, disconnect the receiver under a nitrogen stream and stopper the flask and filter (Figure 8.1d). The solid can now be pumped dry by closing the upper tap on the filter and opening the lower one to vacuum (this prevents finely divided solid flying into the side tube during early stages of drying).
- Finally, the dry solid can be put back into the original reaction flask by inverting the apparatus (swirling the solid around under vacuum will help to remove residues from the walls). Attach a hose from the manifold to the flask, flush it with nitrogen, disconnect the filter under a nitrogen stream and stopper the flask.
- Clean the filter immediately after use and dry it in a hot oven.

As suggested by some of the comments above, finely divided solids can be a nuisance so if you are confident that such a solid is an unwanted by-product (e.g. an alkali metal halide) it can be advantageous to filter through a pad of a filter aid such as Celite, assuming your product will not decompose on the surface of this silicate material. It is best to keep the Celite in a hot oven and when you assemble the apparatus prior to filtration load sufficient onto the sinter to provide a 2–4 cm pad. Arrange the filter vertically then evacuate the apparatus and pump until it is cool. To prevent Celite from entering the side tube, close the upper tap and open the lower tap to vacuum when evacuating, and close the bottom tap and open the top one to nitrogen when admitting the gas. After several purge cycles, open both taps to nitrogen and wash the Celite with dry solvent, allowing the solvent level to reach the surface of the Celite before reducing the pressure slightly below the sinter to compact the Celite pad and then reopen both taps to nitrogen. If any solid is carried through you will have to change the receiver. Since the filter cannot now be inverted and connected to the reaction flask, transfer the mixture into the filter by cannula, taking care to disturb the Celite pad as little as possible. Filter under gravity for several minutes before reducing the pressure below the sinter if it is necessary to speed up the filtration, then wash the Celite pad with fresh solvent unless the filtration was very slow, in which case it might not be worth it.

8.2.2   Removing solvents

To isolate dissolved solids from solution, one option is to concentrate the solution (and if necessary cool it in a freezer) until one component crystallizes. Alternatively, depending on the volatility of the solvent, you might consider removal of all the volatiles in which case solid or involatile liquid products will remain in the flask.

Solvents are most easily evaporated under reduced pressure by attaching the flask to the manifold *via* a cold trap (Figure 3.19). **Never** succumb to the temptation to be lazy and pump volatiles directly into the main cold trap in the line. Contaminants introduced into the hoses on the manifold are likely to find their way into subsequent reactions. Pump out the solvent trap and fill the Dewar flask with liquid nitrogen before opening the tap on the flask and whenever this tap is open agitate the contents by magnetic stirring or by manual swirling to prevent bumping. As the solvent evaporates the contents will cool, so heat the flask in a warm water bath if the warmth of your hands is insufficient to maintain the evaporation rate. This process tends to leave the product spread over the inside of the flask, but it is best to aim to keep as much as possible of the residue in the bottom of the flask. While there is still some solvent present, any material that has splashed onto the walls can be dissolved by closing the tap on the flask and gently warming the contents. This causes solvent to evaporate, condense on the upper walls of the flask and then run back down to the bottom. When you are concentrating a solution prior to crystallization, this is a very effective method of cleaning the inside walls of the flask. However, it is difficult to prevent material splashing up onto the walls when all of the solvent is removed.

When you wish to disconnect the cold trap, close the taps on the flask and the manifold, remove the trap from the liquid nitrogen and ease off the hose from the flask (**do not** twist it). Remove the hoses from the trap and support it in an upright position in a fume cupboard while it warms to room temperature. On a safety note, it is good practice to prepare a label for each solvent and attach the appropriate label(s) to the trap with a loop of wire so that other workers are aware of its contents. Once the trap has reached ambient temperature, consider what it is likely to contain before pouring the liquid into an appropriate waste solvent bottle.

8.2.3 Precipitation

If no crystalline solid is obtained from the solution after concentration and cooling, it may be possible to cause precipitation by adding an excess of another solvent which is miscible with the first but in which the solute is sparingly soluble. This requires some knowledge of the solubility properties of the compound(s) involved and even then it is not always straightforward, and tends to work best with ionic or polar compounds. The dielectric constants of solvents can be used as a guide when choosing suitable solvent combinations but bear in mind the miscibility requirement (note, for example, that polar acetonitrile and non-polar hexane are immiscible). Some combinations that may be appropriate include: addition of hexane to aromatic hydrocarbon solutions; addition of diethyl ether to acetonitrile solutions; addition of diethyl ether to dichloromethane solutions; addition of hexane to dichloromethane solutions. When using this technique, pay close attention to the solution as you add the second solvent since the precipitate may become oily after an initial cloudiness. If this happens you may have to strip off all the volatiles and triturate the residue as described in the next section.

8.2.4   Trituration

Complete removal of the volatiles often results in a sticky or oily product. In such cases it may be possible to obtain a cleaner, free-flowing solid by adding a solvent in which it is sparingly soluble and then scratching with a spatula or glass rod under nitrogen to extract organic impurities. Pull small amounts of the sticky material onto the wall of the flask and press it into a thin film in the solvent. This should break it up into smaller, less sticky particles. Be prepared for this trituration process to take time, effort and patience and when using a glass rod take care not to snap it by applying too much pressure.

After removing the solvent with a dropping pipette, syringe or cannula (it may even be possible to decant it under a nitrogen flow) and repeating the process several times the solid should gradually become less sticky. It is useful to have a second spatula handy to scrape off any material that sticks to the first. Try to do this without removing the ends of the spatulas from the flask in order to avoid exposing too much of your product to the atmosphere. With care this rather tedious and sometimes messy procedure can be quite effective.

8.2.5   Soxhlet extraction

On occasions when the required product is not very soluble and must be separated from insoluble impurities, large volumes of solvents can be avoided by using a Soxhlet extractor (Figure 8.2). The solid mixture is placed in a paper extraction thimble (sintered glass versions are available for moisture-sensitive compounds) and sufficient solvent is added to the flask to prevent it boiling dry during the extraction. The flask is heated to bring the solvent to the boil and the apparatus is adjusted so that solvent drips from the condenser into the extraction thimble where it dissolves some of the product. The solution flows through the wall of the thimble and collects in the extraction chamber until its level reaches the top of the syphon tube. At this point, the solution flows down the inside of this tube, emptying the extraction chamber into the flask by a syphon effect. Multiple extractions are therefore achieved by extended reflux and during this time the product accumulates in the flask. At the end of the extraction, the solvent is removed from the product under reduced pressure.

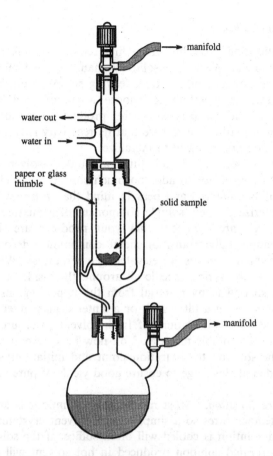

Figure 8.2 Soxhlet extraction under an inert atmosphere.

8.3 Purifying products

Products from the initial isolation procedures described above will invariably be impure and you will then have to decide upon a suitable method of purification before proceeding with characterization. By far the most common method is crystallization and since you will be using this on a routine basis it is crucial that you learn the proper procedures, be aware of the problems you are likely to encounter and know a few tricks to help you overcome them. The basic principles and several variations on the simpler procedures are described in the following section.

8.3.1   Recrystallization

A recrystallization procedure must first remove insoluble impurities and then separate the solute as a solid by changing the temperature, solvent properties or solute concentration, leaving soluble impurities in solution. The first step is therefore to choose a suitable solvent, using your knowledge of the properties of the product to decide upon solvent polarity and volatility and ensure that its reactivity is compatible with the compound you are trying to recrystallize.

Dissolve the solid in the minimum amount of solvent (added from a syringe). Be careful not to add too much solvent in an effort to dissolve any less soluble material and bear in mind that you may want to isolate and characterize this less soluble component of a mixture. Only heat the mixture if you are confident that your products are thermally stable (many organometallic transition-metal compounds decompose or react with chlorinated solvents at elevated temperatures). When adding the solvent, direct the syringe needle all around the inside of the neck of the flask to wash down any material from the upper regions. Filter off any insoluble solid using a filter stick or a sintered glass filter tube and wash the solid with a small amount of fresh solvent to ensure that you have dissolved all the soluble material. You now have several options in order to bring the solution to saturation point and initiate crystallization, and care is needed at this stage to ensure good yields of pure material.

Temperature variation.   Most metalorganic compounds are more soluble at higher temperatures so a simple, single-solvent crystallization in which a saturated solution is cooled will often suffice. If the solute is thermally stable, a saturated solution produced in hot solvent will crystallize as it cools to room temperature, but bear in mind the previous comments about thermal sensitivity. Alternatively, a solution obtained at room temperature can be concentrated to saturation under reduced pressure and then cooled to about −20°C in a freezer. Lower crystallization temperatures can be achieved by using a suitable cold bath (after checking the melting point of the solvent) and this method is useful for compounds that melt near to ambient temperature. It may take several days for crystals to form (this is not likely to be feasible with a cooling bath unless a thermostatted cooling unit is available) and it pays to be patient. If the material forms an oil, re-dissolve it and try to induce crystallization by scratching the flask (on the inside!) with a spatula at low temperature. Slow cooling might also solve this problem so if the solution was obtained with heating, clamp the flask in a hot oil bath and then leave it to cool slowly to room temperature after switching off the heater. If the product remains as an oil, try another procedure.

It is perhaps worth saying something here about 'freezer discipline'.

Since reactions often generate several fractions that may take some time to crystallize you will at some stage find yourself running short of flasks because most of them are in the freezer. It is therefore important not to leave samples in the freezer simply because you aren't sure what to do with them next. Every few days check your samples and decide whether it is worth pursuing their characterization further – there comes a time when it is better to cut your losses and clean out the flask.

To isolate the crystalline material, remove as much as possible of the supernatant liquid (mother liquor) from the crystals at low temperature by using a filter-stick and cannula and then pump off any residual solvent under reduced pressure. Alternatively, after removal of the mother liquor, wash the crystals with a volatile solvent in which they are sparingly soluble (this may be pre-cooled in a separate flask and added *via* cannula) and remove the washings by cannulation prior to drying under vacuum. It is possible to use a sintered glass filter-tube to isolate the solid but if the crystallization has been carried out at low temperature it is more difficult to keep the solution cold during the filtration. Assemble the filtration apparatus, pump it out, flush it with nitrogen and allow it to cool before removing the crystallization flask from the freezer or cooling bath and attaching it to the filter. Use cotton-wool (held with tweezers) to swab the filter-tube with the mixture from a cooling bath, but if the filtration is slow some of the solid is likely to re-dissolve. A cooling jacket can be added to a sintered glass filter but the apparatus is then much less convenient to use since it cannot be inverted.

Solvent modification. When crystals can't be obtained from the chosen solvent, even after cooling to low temperatures, and further concentration leaves an impractically small volume, the solubility of the solute can be reduced by addition of a second, less polar solvent to lower the polarity of the solvent medium. The simplest way to achieve this is to add the second solvent dropwise to the filtered solution at room temperature, swirling the solution after each addition until a faint cloudiness remains. The saturated solution can then be cooled in the freezer until crystals form. When the second solvent is less volatile than the one used to dissolve the solid, it may be possible to precipitate a crystalline solid by controlled evaporation of the more volatile component under reduced pressure. To grow larger crystals for X-ray diffraction studies a slower evaporation rate can be achieved by passing nitrogen over the solution (with an oil bubbler attached to monitor the flow rate). This approach is often used for air-stable compounds where the solution can be left to evaporate slowly on the bench top or in a fume cupboard.

A convenient way to change the composition of the solution is to layer a less dense, less polar solvent on top of the filtered solution. This should be done with care to avoid precipitating the solid as a fine powder. Add the second solvent **slowly** from a syringe (or *via* a cannula from another flask) by pressing the tip of the needle against the wall of the flask slightly above the meniscus and running the solvent down the wall, taking care not to allow drops to fall into the solution. The interface will initially become cloudy but gradually a layer of clear solvent should build up above the solution. Diffusion is surprisingly slow and crystal growth, which initially occurs near the interface, can take from several days to several weeks during which time the cloudiness will disappear, but don't despair if at first an oil forms, this may well crystallize on prolonged standing. Check to see whether diffusion is complete by swirling the flask gently and looking for density gradients in the liquid; if they are absent, the mother liquor can be removed.

For those compounds that are difficult to crystallize, vapour diffusion is well worth a try. Under a nitrogen atmosphere, the flask containing the filtered solution is connected to one containing a volatile solvent in which the compound is less soluble (Figure 8.3). Alternatively, an integrated design containing a sinter (Figures 3.18a and 3.18b) can be used. Carefully and briefly open a tap to vacuum to reduce the pressure in the system taking care not to allow the solution to bump. With all the taps closed, the less polar solvent will begin to diffuse into the solution. Again, this can be a slow process, so be patient!

Whichever of these solvent modification processes is used, when you are satisfied that most of the solid has crystallized remove the mother liquor with a syringe or by cannulation and wash the crystals with a suitable solvent before pumping them dry. If the crystals are for an X-ray crystal structure determination, it may be safer to leave some of the mother liquor in the flask in case they lose solvent of crystallization on drying. If the solid is finely divided and it is necessary to use a filter stick (section

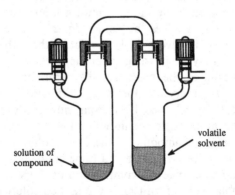

volatile solvent

solution of compound

Figure 8.3 Crystallization by vapour diffusion.

3.4.2), be wary of introducing paper fibres into the sample, they can play havoc with microanalysis results.

Crystals for X-ray diffraction studies are often obtained from small-scale crystallization and specific variations on the methods discussed above are described in section 11.5.

8.3.2 Distillation

Liquids are most readily separated and purified by distillation and you should be familiar with the proper procedures and be aware of the available options, even if you are only likely to need them in the preparation of ligands and reagents.

Separation by distillation depends upon the difference in vapour pressures of the components of the mixture and each stage of vaporization and subsequent condensation enriches the condensate in the more volatile component. Simple distillation is essentially a single stage process and is only effective if the components differ in their boiling points by at least 80°C; otherwise it is necessary to use a fractionating column in which a contraflow of vapour and condensed phases is equivalent to multiple vaporization–condensation stages. The complete set of apparatus required for distillation therefore includes a distillation flask and some means of heating it, a fractionating column (although this is not always necessary), a still head, a condenser (this is often incorporated into the still head) and some means of collecting the distillate. Incorporation of the still pot and fractionating column into a single unit (Figure 8.4) eliminates the need for a joint which might otherwise be submerged in heating oil or make it difficult to clamp the glassware. It is important to use

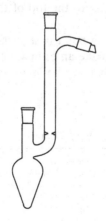

Figure 8.4 Distillation flask.

apparatus of a size appropriate for the amount of material to be distilled (use a flask size that is about 1.5 times the sample volume) and since this usually means having a range of flask and column sizes available, glassware for distillation tends to be communal.

Fractionating columns are most often of the Vigreux type or else can be packed with glass beads, rings or helices, and while Vigreux columns are less efficient, they have the advantage of having lower hold-up volumes than packed columns, although an alternative wire gauze packing which reduces this problem has been described by Loewenthal [2]. Fractionating columns work best when lagged, and insulation is improved by incorporation of a vacuum jacket. The efficiency of a fractionating column is related to its length, and the length of a column equivalent to one simple distillation is referred to as one theoretical plate. Table 8.1 compares the efficiencies of different designs of fractionating columns [3].

Spinning-band and 'Spaltrohr' columns are high efficiency, low hold-up designs which are suitable for separation of small volumes of compounds having boiling points as close as 3°C. The former contains a spiral Teflon band rotating at high speed within the column while the latter consists of concentric grooved tubes. Both are commercially available and are relatively expensive, while the other designs mentioned above can be constructed by a glassblower.

During fractional distillation, the apparatus should allow the reflux ratio (i.e. how much you collect compared with how much runs back down the column from the condenser) to be controlled. A versatile partial take-off still head for inert atmosphere distillation is shown in Figure 8.5 attached to the top of a fractionating column, although it can also be connected directly to a Schlenk flask. The long heating jacket shown in Figure 8.6 protects the column from draughts and allows heating oil to be added right up to the top of the column for high boiling materials [2].

If you don't know the boiling point of a particular compound and can't find it in the literature, then make an estimate by comparison with that of a closely related molecule. When the boiling point at atmospheric pressure

Table 8.1 Efficiencies of fractionating columns

Column type	Diameter (mm)	Theoretical plate height (mm)
Empty column	6	150
Vigreux	12	77
Packed column	24	60
Spinning band	5	25

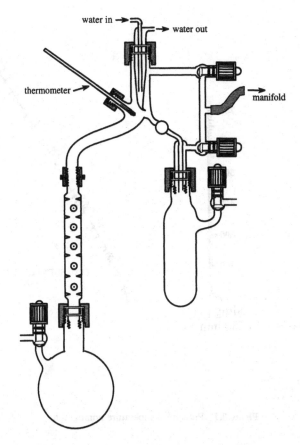

Figure 8.5 Fractional distillation under an inert atmosphere.

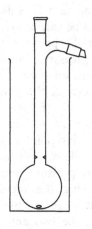

Figure 8.6 Extended heating bath.

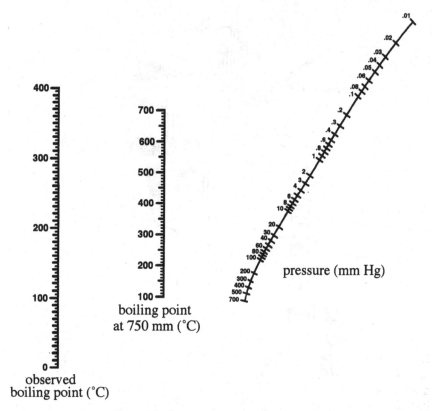

Figure 8.7   Pressure–temperature nomograph.

is beyond the range of a heating bath or is likely to result in some decomposition, the distillation should be carried out at a reduced pressure. As a rough rule of thumb, halving the pressure reduces the boiling point by 10°C, so an estimate of the pressure required for a convenient boiling point can be derived from a known boiling point at a different pressure. Alternatively, by pivoting a ruler at the atmospheric boiling point on a pressure–temperature nomograph (Figure 8.7), temperature–pressure combinations can be read off. The reverse process provides the atmospheric boiling point from a boiling point at some other pressure.

The pressure in the vacuum side of your manifold will often be too low for distillations under reduced pressure and in these cases a manostat provides a means of controlling the pressure between about 5 and 50 mmHg. A typical manostat arrangement has a mercury float valve in the vacuum line (Figure 8.8). The two anti-surge bulbs smooth out the fluctuations in pressure and should be enclosed in plastic mesh.

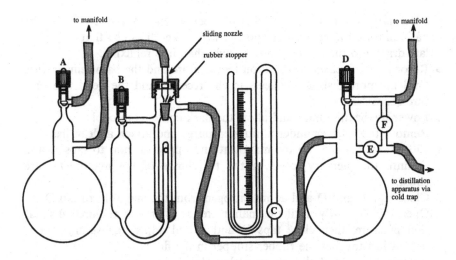

Figure 8.8 Manostat apparatus.

The general procedure for a reduced pressure, inert atmosphere distillation is described below, including details of manostat operation. For a distillation that can be carried out at atmospheric pressure or one which does not require a manostat, omit the relevant sections from the instructions. It is important to first assess the volatilities of the various components of the mixture to be distilled. The following procedure includes the removal of volatile compounds at atmospheric pressure before distillation of the remainder under reduced pressure.

- Set up a fractionating column, partial take-off head and receiver (Figure 8.5) and connect to a manostat (Figure 8.8) *via* a solvent trap (Figure 3.19). Connect taps A and D of the manostat to the manifold.
- Cap the end of the fractionating column, open all the taps (including A–F on the manostat) evacuate the whole apparatus and flush with nitrogen.
- Connect the distillation flask (containing the mixture to be distilled and a magnetic stirrer bar) to the fractionating column under nitrogen.
- Lag the column with glass wool or other suitable insulating material and turn on the water supply to the condenser. Close the take-off tap on the distillation head.
- Immerse the distillation flask in a heating bath and stir the contents while heating the bath steadily. Clamp a thermometer in the bath to monitor the temperature.
- Adjust the heating rate so that the low boiling fraction refluxes at the

condenser at a rate of several drops per second. Make a note of the bath and head temperatures and open the take-off tap so that the distillate drips into the receiver more slowly than it is refluxing.

- Record the increasing distillation temperature and the bath temperature until no more distillate collects in the receiver and the head temperature decreases.
- Lower the heating bath and allow the flask contents to cool.
- Remove the receiver under a nitrogen purge and attach a clean flask.
- Close taps E and F on the manostat and apply vacuum to taps A and D until the mercury levels in the two limbs of the manometer are equal.
- Close taps A and D and admit nitrogen from the manifold to tap D.
- Open tap D slowly until the manometer registers the required distillation pressure, then close taps D and B and adjust the sliding nozzle until it just touches the rubber stopper in the float.
- Reopen tap A and the pressure should remain constant. If there is a slow decrease in pressure, open tap D slightly to allow a very slow nitrogen bleed into the manostat.
- When the contents of the distillation flask have cooled, immerse the solvent trap in liquid nitrogen and open tap E. The distillation apparatus should now be evacuated to the set pressure.
- Distil the remaining liquid as previously, increasing the temperature of the bath slowly and steadily and changing receivers if several fractions need to be collected. This can be done by closing the uppermost tap on the distillation head and tap E on the manostat, while opening tap F to allow a nitrogen purge when the receiver is changed. To re-establish the distillation pressure, close tap F and then open tap E and the uppermost tap of the distillation head.

Do not try to distil to dryness to maximize your yield, since decomposition is likely to occur on overheating and, in any case, there will always be some liquid that will fall back into the distillation flask when the heating is stopped (this represents the column hold-up).

When only a small amount of material is being purified, specially designed glassware can prevent losses due to hold-up that occur in conventional apparatus and examples are shown in Figure 8.9. Problems often arise with the design shown in Figure 8.9a because of the proximity of the receiver to the heating bath, and a neat alternative is the vertical arrangement shown in Figure 8.9b which is easily modified for individual requirements. The joint which accommodates a thermometer allows easy removal of the distillate under nitrogen with a syringe. Bulb-to-bulb distillation using a Kugelrohr (Figure 8.10) is common in organic laboratories and this technique may be useful for the purification of high boiling ligands. Some fractionation is possible with this apparatus by immersing

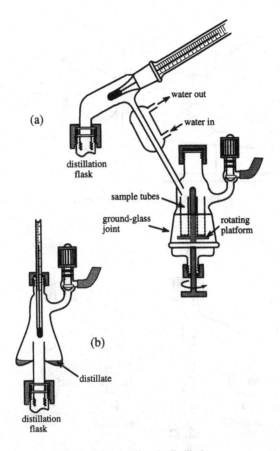

(a)

water out

water in

distillation
flask

sample tubes

ground-glass
joint

rotating
platform

(b)

distillate

distillation
flask

Figure 8.9 Small-scale distillation.

all except the last bulb (i.e. the one nearest the vacuum connection) in the oven and withdrawing successive bulbs as the temperature increases. The distillate is removed from the bulbs by pipette and, although this can be done under a nitrogen purge, the technique is not well suited to very air-sensitive compounds.

One particular type of distillation that could, in principle, have been included in chapter 7 is that involving a Dean–Stark trap. The designs shown in Figure 8.11 are used to remove water from an equilibrium and thereby drive the reaction to completion. In the somewhat cumbersome traditional design (Figure 8.11a), an immiscible solvent is used (e.g. toluene) which forms an azeotrope with water so that when the azeotrope is collected the water can be removed as a separate phase. However, the smaller, neater adapter shown in Figure 8.11b, which uses molecular

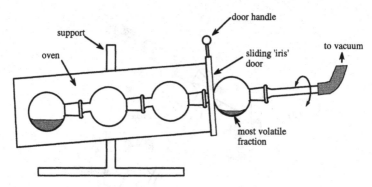

Figure 8.10 Use of a Kugelrohr.

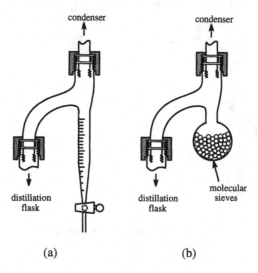

Figure 8.11 Dean and Stark traps.

sieves to absorb water, is more efficient and can be used for azeotropic distillations with water-miscible solvents.

8.3.3 Sublimation

Many inorganic and metalorganic solids are sufficiently volatile to be purified by sublimation, although it is often difficult to avoid some material losses due to decomposition. Some thought must also be given to the best method of collecting the purified product. Volatile components of a mixture can be separated by fractional sublimation in a

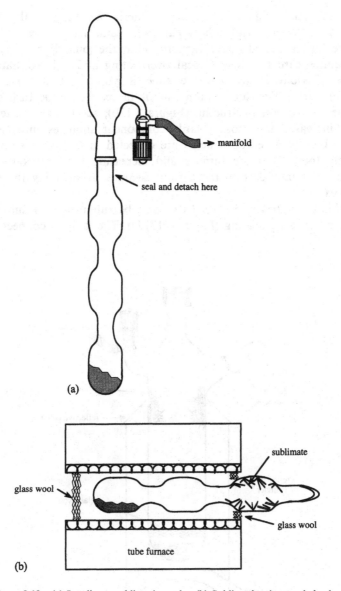

Figure 8.12 (a) Loading a sublimation tube. (b) Sublimation in a sealed tube.

sealed tube, which should be made with a joint so that it can be
evacuated, flame dried and then connected to the flask containing the
sample. The whole assembly is then evacuated and the solid is tipped
into the sublimation tube with gentle tapping so as to transfer it to the

end of the tube (Figure 8.12a). Any material adhering to the walls is sublimed to the end with a heat gun or a flame from a gas torch and then the tube is sealed under vacuum below the joint. For larger tubes, this requires care to prevent local overheating and hole formation, so ask the glassblower to seal the tube if you don't feel sufficiently confident. The sealed sublimation tube is placed in a tube furnace with only the last section protruding (Figure 8.12b), so that as the temperature is increased the most volatile component sublimes into the cool portion. Less volatile components are collected in the other sections by removing these from the furnace and increasing the temperature. The sublimate is removed from the various sections by breaking the tube in a dry box.

Sizeable quantities of material can also be sublimed in a sublimation flask using infrared heating (Figure 8.13) [4]. The lamp is connected to a

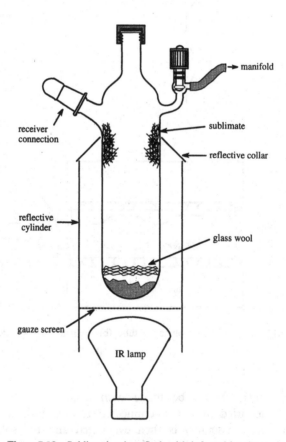

Figure 8.13 Sublimation in a flask with infrared heating.

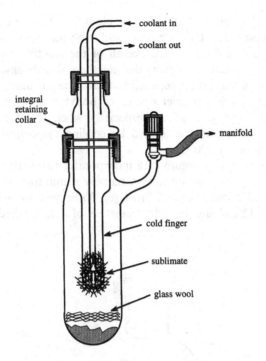

coolant in

coolant out

integral
retaining
collar

manifold

cold finger

sublimate

glass wool

Figure 8.14 Sublimation onto a cold finger.

variable power source and the voltage increased gradually until sublimation commences. We find this technique to be very convenient and use it
to sublime up to 20 g of WOCl$_4$ at a time on a high vacuum line using
large sublimation flasks (60 mm diameter, 400 mm long). The sublimate is
scraped out under nitrogen into a flask connected to the side-arm.

Other sublimer designs usually employ a cold finger (Figure 8.14) which
can be cooled with running water or solid CO$_2$. Air-sensitive compounds
must be removed from this type of apparatus in a dry box and it may be
difficult to prevent the sublimate from dropping back into the residue
during the dismantling process.

8.3.4 Chromatography

The difficulties associated with thin layer chromatography have already
been briefly mentioned in section 7.7. Similarly, problems associated
with the decomposition of metalorganic/organometallic compounds on
silica gel or alumina coupled with the expense of the high-grade
supports and the large volumes of solvent required mean that in
inorganic laboratories column chromatography is not the universal

technique that it has become in organic laboratories. For work with sensitive compounds, the support and solvent must be rigorously degassed and it is difficult to remove all of the reactive surface hydroxyl groups on the support. However, despite these drawbacks column chromatography is an important separation technique for many organometallic compounds. Cyclopentadienyl and carbonyl complexes in particular can often be eluted successfully, especially when the decomposition of more sensitive compounds is limited by cooling the column to temperatures in the region of −70°C.

The column shown in Figure 8.15 incorporates two features which have been described to increase efficiency, namely a multiple bore [2,5] and a vacuum-jacketed cooling arrangement [6]. Tube sections with lengths of approximately 12 cm and outside diameters of 1.2, 1.7 and 2.4 cm have

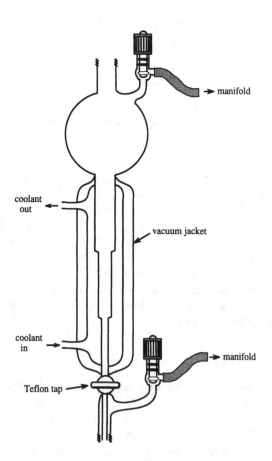

Figure 8.15 Jacketed chromatography column for air-sensitive compounds.

been recommended for between 5 and 40 g of adsorbent, which may be alumina (this may have to be deactivated by treatment with up to 15% of water), silica or Florisil (a magnesium silicate which is less active than alumina) and is loaded into the column as a slurry in a solvent such as hexane. The slurry can be made up in a separating funnel with a Teflon tap from which it is run directly into the column. Tap the column with a cork ring or piece of rubber hose to ensure an even packing as the adsorbent settles and run off the solvent to leave 2–3 mm of solvent above the surface, which should be as even as possible. Add a solution of the sample (which should be as concentrated as possible) by carefully running it down the wall of the column so as not to disturb the surface of the adsorbent (a piece of filter paper cut to size and placed on top of the column can also help). Allow this solution to run into the column, leaving 2–3 mm above the surface, then add a small volume of eluent and repeat this exercise. Add the main volume of eluent to the reservoir and elute the column at a rate which is convenient and gives a good separation. Preliminary TLC on the mixture will determine the most appropriate eluents, and typical solvents include hexane, toluene, diethyl ether, tetrahydrofuran, ethyl acetate, dichloromethane and acetone. It is straightforward to collect coloured compounds as they come off the column, and colourless compounds that fluoresce can be detected by shining a portable UV lamp onto the column (**Care!**); otherwise fractions must be collected and analysed spectroscopically (UV/Vis or NMR) to determine which ones contain the various components of the mixture.

Variations on this technique that offer some advantages include flash chromatography and dry column chromatography and these are described in detail in reference 1.

8.4 Storing products

When you have isolated, purified and characterized a compound it is impractical to leave it in a flask for any length of time – you would soon have no flasks for new reactions. Air-sensitive compounds cannot be kept in the usual types of sample tubes with plastic push-fit caps or screw tops unless they are stored inside a dry box. Ensure that sample tubes and their caps are thoroughly dried before taking them into the box and don't use screw tops with laminated card inserts since they retain moisture and are difficult to dry. It is important to label each tube with the sample number from your lab book (which should incorporate your initials to make it obvious to others whose sample it is) and it is also useful to include the date. However, dry boxes soon become congested so the best method of storage is to seal your samples in glass ampoules since these can be kept outside the dry box.

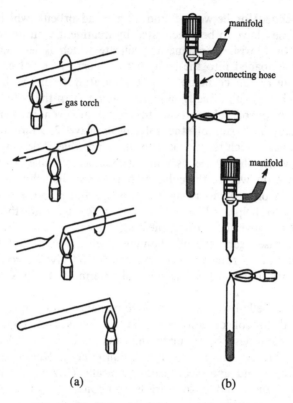

Figure 8.16   Construction of small glass ampoules and sealing a sample under vacuum.

8.4.1   Ampoules

Simple ampoules that can hold up to about 0.5 g are easily made from small-bore glass tubing. Select tubing with a diameter that can be connected *via* a short length of rubber or PVC hose to a suitable vacuum tap and then proceed as below.

- Cut the tubing to length with a glass knife or file.
- Seal off one end of the tubing in the flame from a gas torch, making sure that there are no leaks (blow gently into the tube while the end is softened to keep it rounded), and then fire polish the other end to remove sharp edges (Figure 8.16a).
- Allow the glass to cool then rinse with ethanol and dry in a hot oven.
- Label the ampoule with a marker pen above and below where you intend to seal it off.
- If you don't have a balance in the dry box, weigh the ampoule while it is still hot.

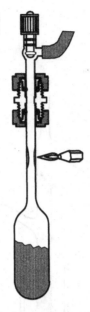

Figure 8.17   Sealing larger glass ampoules under vacuum.

- Take the ampoule into the dry box with the sample, a vacuum tap, a short piece of connecting hose (2–3 cm) and a small glass funnel (the transfer of solids is easier if you attach a funnel to the neck of the ampoule with the connecting hose).
- If you have a balance in the dry box, weigh the empty ampoule.
- Attach the funnel, load the sample into the ampoule (but don't fill it much more than half full) and then detach the funnel.
- If there is a balance in the box, weigh the filled ampoule.
- Attach the vacuum tap, close it and remove the assembly from the dry box.
- Attach the tap to a manifold and pump out the hose from the vacuum line.
- Slowly open the tap to evacuate the ampoule, taking care not to allow solid to be pumped into the line.
- Seal the ampoule with a gas torch by heating sufficiently far above the sample to prevent thermal decomposition and far enough below the tap to avoid melting the connecting hose. Use a small flame and heat opposite sides of the tube to give four inward-facing indentations then heat this whole region with a larger flame to melt the glass, pulling the ampoule slightly to ensure a good seal. With continued heating, keep pulling until the ampoule is detached then melt the tip of the seal and allow to cool (Figure 8.16b).

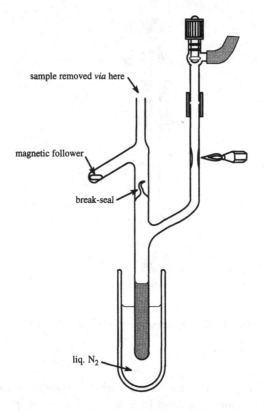

sample removed *via* here

magnetic follower

break-seal

liq. N$_2$

Figure 8.18 Sealing a volatile sample in a break-seal ampoule.

- If there was no balance in the dry box, re-weigh the ampoule and the detached glass to determine the weight of sample.

These small ampoules are easily opened in the dry box or in an inert gas stream by breaking the glass. First score the tubing with a glass knife for about half of its circumference (do not try to saw the glass), then grasp it at either side of the scratch and bend it while pulling the two ends apart. In the dry box, you can protect against glove punctures as a result of unclean breaks by wrapping the scored tube in paraffin film.

Unless you are adept at glassblowing, larger ampoules are best prepared by the glassblower. A wider neck facilitates sample loading and a constriction in the neck makes it easier to seal the ampoule. We have the necks of our larger ampoules made with an outside diameter to match that of 10 mm internal diameter Young's taps (POR 10RA, o.d. *ca.* 12.5 mm). This enables us to connect the tap to the ampoule using O-ring

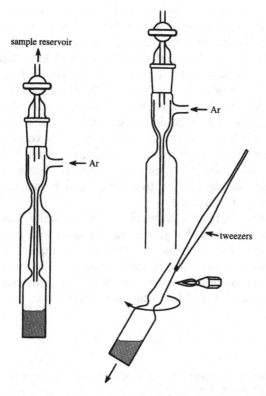

Figure 8.19 Filling and sealing commercial ampoules under an inert atmosphere.

compression couplings made by modifying standard commercial plumbing fittings (Figure 8.17). More care is needed when opening larger ampoules because of the greater thickness of the glass. Before scratching the tube, make sure that the glass knife is sharp and wrap the tube in paraffin film since it is more likely that glass fragments will be produced than with small diameter tubing. An alternative method of breaking the glass tube is to heat the end of a glass rod to red heat in a hot flame and apply it to the glass just beyond the end of the scratch. A crack should develop and follow the scratch. Some liquids with boiling points close to room temperature are supplied in ampoules of up to 500 cm^3 and these should be cooled carefully (to avoid cracking the ampoule) in ice-water or another cooling bath to reduce the vapour pressure before opening the ampoule in a fume cupboard.

Volatile compounds can also be stored in ampoules, but in this case a break-seal design allows the sample to be transferred into and out of the ampoule on the high-vacuum line. With a little practice, these can be made from glass tubing, but if you lack the skills or the confidence have them made by the glassblower. A typical design is shown in Figure 8.18 which also depicts sample transfer on the vacuum line. After heating the ampoule under vacuum on the line, the volatile sample is condensed in at low temperature as described in section 5.3. The ampoule is then sealed off at the point indicated and allowed to warm to room temperature. To remove the sample, the ampoule is connected to the vacuum line with a small magnetic follower in the leg as shown. After cooling the ampoule and evacuating the connecting tube, a magnet is used to transfer the follower into the tube with sufficient force to break the glass seal, whereupon the sample can be transferred under vacuum into another flask by warming the ampoule.

The ampoules supplied commercially are generally less amenable to inert atmosphere work, although they can be filled with liquids by using a purging adapter (Figure 8.19) and then sealed quickly in a flame by holding the glass of the neck in metal tweezers. This method has been used to store organophosphines in convenient weighed amounts so that a whole ampoule can be used for each reaction [7].

References

1. Casey,M., Leonard, J., Lygo, B. and Procter, G. (1990) *Advanced Practical Organic Chemistry*, Blackie, Glasgow.
2. Loewenthal, H.J.E. (1990) *A Guide for the Perplexed Organic Experimentalist*, 2nd edn, John Wiley and Sons, Chichester.
3. Keese, R., Müller, R.K. and Toube, T.P. (1982) *Fundamentals of Preparative Organic Chemistry*, Ellis Horwood Ltd., Chichester.
4. Pool, G.C. and Teuben, J.H. (1987) In *Experimental Organometallic Chemistry: a Practicum in Synthesis and Characterisation*, eds A.L. Wayda and M.Y. Darensbourg, ACS Symposium Series, Vol. 357, p. 30.
5. Fisher, G.A. and Kabara, J.A. (1964) *Anal. Biochem.*, **9**, 303.
6. Buck, R.C. and Brookhart, M.S. (1987) In *Experimental Organometallic Chemistry: a Practicum in Synthesis and Characterisation*, eds A.L. Wayda and M.Y. Darensbourg, ACS Symposium Series, Vol. 357, p. 27.
7. Adams, D.M. and Raynor, J.B. (1965) *Advanced Practical Inorganic Chemistry*, John Wiley and Sons, London.

9 Reactions Between a Solid and a Gas

9.1 Introduction

You are most likely to encounter high temperature reactions where a gaseous reagent is passed over a heated solid (usually a metal or metal oxide) in the preparation of metal halides for use as starting materials. In the following sections, some general considerations are discussed and illustrated with reference to specific examples. Section 6.5 should be consulted for the handling of gaseous reagents.

9.2 Designing apparatus

The first considerations when carrying out a reaction of this type must be the physical properties of the reagents and products. Gaseous reagents can be used directly or diluted in an inert gas stream, while volatile liquid reagents are usually entrained in an inert gas stream. Impurities in the gas must be minimized because of the significant volumes exposed to the reactive solid, especially when a carrier gas is being used. Reagent gases will either be used directly from a cylinder or generated as required (section 6.5) while an inert carrier gas can be purified as described in section 3.2.5. In the majority of cases, volatile products can be transferred from the reactor by sublimation and, if they are air-sensitive, provision must be made for their isolation under an inert atmosphere. The apparatus should also include a scrubber stage to remove unreacted, hazardous, gaseous reagents.

9.2.1 Horizontal reactors

The most common basic arrangement of apparatus for this type of reaction is shown in Figure 9.1 [1]. The manifold inlet arrangement enables a flow of either inert gas, reagent gas or a mixture of the two to be passed through the apparatus, while a bubbler is incorporated for drying agents such as concentrated sulphuric acid. An additional gas inlet may be incorporated as shown to allow pre-reduction of a metal with hydrogen. The reaction tube itself can be of borosilicate glass. However, if temperatures above 500°C are required or if the reaction is highly exothermic, a silica tube should be used for the heated portion. Graded

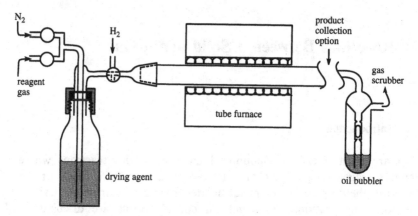

Figure 9.1   Horizontal reactor for solid–gas reactions.

silica/borosilicate seals are available to enable the addition of borosilicate sections or joints to the ends of a silica reaction tube. Volatile products can be sublimed from the reaction tube, and in these cases the solid reactant can be placed directly into the reaction tube. Non-volatile solids are more difficult to remove after the reaction, and in these cases the solid is placed in one or more porcelain boats which are pushed into the reaction tube. The product is then retained within the boats. What you attach to the exit from the reaction tube will depend on how you intend to collect the product. Integral glass bulbs (Figure 9.2) can be used for subsequent purification, but these must be sealed off with a gas torch and opened in a dry box. Alternatively, the product can be collected in a Schlenk flask to enable straightforward isolation under an inert atmosphere (Figure 9.3). Porcelain boats containing non-volatile products can be pushed into a Schlenk flask connected directly onto the end of the reaction tube.

Tube furnaces are often constructed in the workshop to provide the temperatures required, although commercial designs are also available (but more expensive). In either case, a provision for temperature measurement (with an integral thermocouple or a thermocouple probe) must be included.

When an entrained vapour (e.g. Br_2 or CCl_4) is required, the arrange-

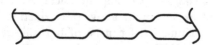

Figure 9.2   Product collection bulbs.

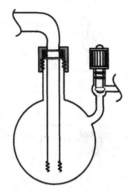

Figure 9.3 Arrangement for collecting product in a Schlenk flask.

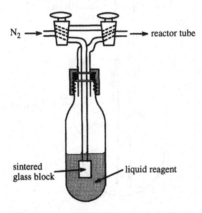

Figure 9.4 Entrainment of a volatile liquid reagent in a carrier gas.

ment shown in Figure 9.4 can be used. This incorporates a by-pass for isolation of the flask containing the reagent.

9.2.2 Vertical reactors

Gravity assists in the removal of heavy product vapours from a vertical reaction tube and this arrangement has been recommended for the preparation of VCl_4, $NbCl_5$, $TaCl_5$, WBr_5 and $ReCl_5$ from the metals and the respective halides [2]. I can vouch for the convenience of this method in the preparation of $ReCl_5$, and Figure 9.5 shows a slightly modified version of the apparatus described in ref. 2. The nitrogen, hydrogen and halogen are admitted through a gas inlet manifold connected to the top of the reaction tube, while the metal is supported on quartz wool seated

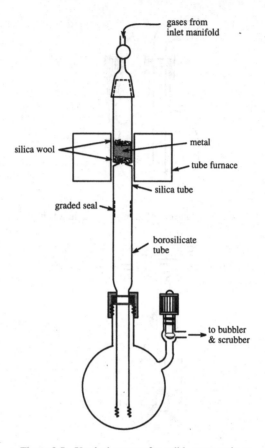

Figure 9.5 Vertical reactor for solid–gas reactions.

on indentations in the silica tube. A small furnace is used to heat the region around the metal powder.

9.3 Procedure

A reaction of this type should be carried out in a fume cupboard following the general procedure outlined below.

- Decide on the apparatus to be used and assemble it from the drying oven.
- For a vertical reactor, insert a wad of dried silica wool (we bake it in an oven, place it in a flask and pump under vacuum) onto the indentations in the tube and then carefully place the weighed metal powder

onto this support. Another wad of silica wool can be inserted above the metal.

- For horizontal reactors, either place the solid reactant directly into the tube using a scoop-type spatula (the handle can be extended by pushing a piece of rubber or plastic tubing over one end) or put it into one or more porcelain boats and insert these into the tube.

- Flush out the apparatus with inert gas.

- To remove surface oxide from a metal prior to halogenation, pass hydrogen through the system (or a mixture of hydrogen and inert gas) and heat the metal to about 300°C. When no more water is formed, replace the hydrogen by oxygen-free nitrogen and flame out the reaction tube and receiver to remove the water (if the apparatus can be connected to a vacuum line, careful evacuation with the bubblers isolated will aid this process).

- Heat the solid to the required temperature and admit the reagent gas, diluting with nitrogen if the product forms too quickly. If an entrained vapour is being used, the by-pass valves should be turned to pass nitrogen through the reagent liquid, which can be warmed in a heating bath if necessary.

- (Note that if CCl_4 is being reacted with a metal oxide, the phosgene ($COCl_2$) formed is toxic and should be passed into an alkali scrubber after the exit bubbler.)

- Control the flow rate to prevent a volatile product from being swept out of the apparatus in the gas stream. In cases where the product is a liquid, cool the receiver in an ice bath.

- When the reaction appears to be complete, replace the reagent gas with inert gas and use a flame or a heat gun to sublime any volatile product out of the reaction tube and into the receiver flask or the first glass bulb. When all of the product has been sublimed into the bottom of the flask, remove and stopper it under a flow of nitrogen.

- With glass bulb receivers, the product can be further purified by bulb-to-bulb sublimation by moving the furnace slowly along the tube. The bulb containing the pure product is sealed off with a gas torch at the constrictions, opened in a dry box and scraped out into a suitable storage container.

- If the product is non-volatile and is contained within a porcelain boat, establish a nitrogen flow through a receiver flask connected directly to the reaction tube, remove the inlet joint from the tube and push the boat into the flask with an extended spatula or glass rod. Remove the flask from the tube, tap the flask to tip out the solid from the boat and use a piece of wire with a bent end to remove the boat from the flask. Cap the flask.

- For extended storage, seal the product in labelled, tared, glass ampoules that can be kept outside the dry box.

This procedure is applicable to a range of transition metal and main group halides as well as some other binary and ternary compounds (see chapter 13).

References

1. Angelici, R.J. (1977) *Synthesis and Technique in Inorganic Chemistry*, W.B. Saunders, Philadelphia, pp. 33–38 and references therein.
2. Lincoln, R. and Wilkinson, G. (1980) *Inorg. Synth.*, **20**, 41.

10 Reactions Between Solids

10.1 Introduction

In addition to the synthesis of ever larger polynuclear molecules, the excision of metal clusters from solid-state materials serves to illustrate the increasing overlap between disciplines traditionally regarded separately as 'solid-state' and 'molecular' inorganic chemistry. It is therefore evident that some knowledge of the solid-state chemist's art is likely to be beneficial to those of us who are more familiar with molecular compounds, and this chapter has been included to establish some familiarity with solid–solid reactions relevant to the preparation of potentially useful starting materials for metalorganic chemistry. I will assume here that a working knowledge sufficient to enable you to carry out literature procedures will prove adequate, but for those who aspire to more investigative reactions, a review by Corbett provides a good introduction to solid-state synthesis [1].

You are most likely to encounter reactions between solids when preparing the lower oxidation state halides of the transition metals and, given the developments in molecular cluster chemistry based on solids containing hexanuclear cluster units, I have included in the following sections descriptions of the high-temperature preparations of Mo_6Cl_{12}, W_6Cl_{12} and related rhenium compounds to illustrate the general procedures involved in solid-state synthesis.

10.2 General considerations

The low rates of diffusion or mass transfer associated with mixtures of solids mean that high temperatures are usually required to achieve reasonable rates of reaction, although chemical (vapour phase) transport enables lower temperatures to be used than those required for volatilization of the reagents themselves. In this technique, a transporting agent acts as a 'gaseous solvent' by transforming non-volatile reactants and products into volatile derivatives which can then interact. In investigative solid-state reactions, it is frequently the case that single-phase, homogeneous products are not obtained and crystalline products must be separated by hand. However, yields in established reactions used for the preparation of starting materials have usually been optimized and this problem should not arise. Reactions can be carried out under an inert gas flow or in sealed reaction vessels, usually in a tube furnace with some provision for moving the heated zone.

10.3 Choice of container

The design of the reaction vessel will depend on whether a gas flow or a sealed system is to be used, and the material of choice will most likely be borosilicate or silica glass, although in some cases a refractory metal such as tantalum must be used. Horizontal reaction tubes as described in chapter 9 are suitable when an inert gas flow is required, as in the preparation of Mo_6Cl_{12} using a silica tube (Figure 10.1). Pressures can develop in sealed reaction tubes due to volatile starting materials or products, so these tubes are often sealed under vacuum, if necessary with cooling to avoid loss of volatile starting materials. This presents no major problems with borosilicate glass, but silica requires a higher temperature flame (e.g. hydrogen/oxygen) and you may have to ask a glassblower to seal these tubes for you. Reactions involving volatile reagents (or products) such as sulphur, bromine and iodine can be safely carried out without developing dangerous pressures as long as the tube is well evacuated and the volatile reagent (or product) can move to a cooler zone. Typical designs for sealed-tube reactors are shown in Figure 10.2.

Refractory metal containers are necessary for high temperature reactions involving reactive transition and rare earth metals, since these elements will react with glass to form very stable oxides. Tantalum has generally been found to be the most suitable metal for synthetic applications, although niobium, molybdenum and tungsten may also be used when necessary. The techniques associated with fabrication, welding and use of metal containers have been described [2], and Figure 10.3 shows some typical designs which can be constructed from high-purity tubing with a wall thickness of 0.25–0.76 mm and a diameter of 6–9 mm.

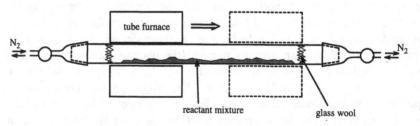

Figure 10.1 Horizontal reactor for solid-state reactions.

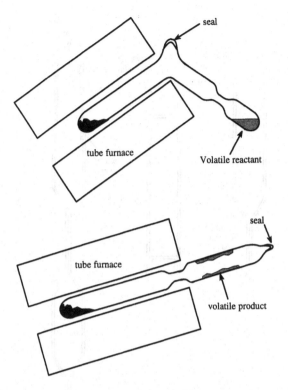

seal

tube furnace

Volatile reactant

seal

tube furnace

volatile product

Figure 10.2   Solid-state reactions in sealed glass tubes.

10.4   Typical procedures

10.4.1   Glass containers

In many reactions, the solids can be mixed directly (in a dry box if they are air-sensitive) and then transferred to the reaction tube. The transfer to a long horizontal flow reactor will have to be done outside the dry box and may require some ingenuity. As an illustration, the procedure used in the preparation of Mo_6Cl_{12} is described here [3].

- Place a glass wool plug in one end of a silica tube fitted with taps as shown in Figure 10.1.
- Dry the apparatus by heating the tube while a slow stream of purified nitrogen is passed through it, then allow to cool.
- In a dry box, add the correct amounts of $MoCl_5$ and molybdenum powder to a Schlenk tube and mix thoroughly. Then bring the closed tube out of the box.
- Under a stream of nitrogen, connect the Schlenk flask to the end of the

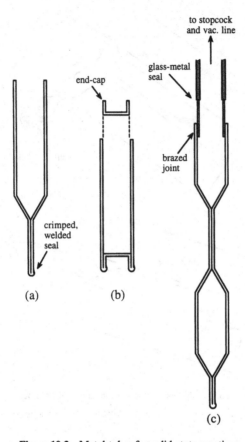

Figure 10.3   Metal tubes for solid-state reactions.

reaction tube without the glass wool plug and gently tip the solid into the reactor, tapping the walls of the flask to ensure complete transfer.

- Remove the Schlenk flask, add a dry glass wool plug and replace the tap on the end of the tube.
- Distribute the mixture evenly along the length of the tube and move the furnace to heat the solid nearest the nitrogen inlet while a slow stream of nitrogen is passed through the apparatus.
- Heat to 600–650°C, and when the solid is yellow–brown, slowly move the furnace along the tube until the rest of the reaction mixture is the same colour.
- The volatile halides will precede the furnace as it moves, and when they reach the glass wool plug, stop heating.
- Allow to cool a little, reverse the flow of nitrogen and resume heating, moving the furnace in the opposite direction.

- Repeat these heating cycles until a negligible amount of volatile halides is visible (about five cycles) and allow the apparatus to cool.
- Remove any volatile materials that may have collected at both ends and transfer the remaining mass of yellow solid to a beaker for work-up in aqueous 25% HCl.

Sealed tube reactors are likely to be small enough to be loaded with the reagents in a dry box, after which they are capped with a tap adapter, removed from the dry box and sealed under vacuum. For example, in the synthesis of W_6Cl_{12} from WCl_4 and iron powder [4], the solids are mixed and loaded into a borosilicate tube which is subsequently sealed under vacuum and placed in a horizontal tube furnace. After being heated slowly to 500°C and held at that temperature for 3 days, the tube is allowed to cool and then re-positioned so that the sample remains in the furnace with the opposite end projecting out. After maintaining a temperature of 450°C for several hours to condense volatile materials in the cooler zone, the tube is allowed to cool and then broken open in air for subsequent work-up.

In a variation on this method, $Rb[Re_6S_5Cl_9]$ and related hexanuclear rhenium cluster compounds have been synthesized by reduction of $ReCl_5$ with rhenium metal in the presence of sulphur and rubidium chloride [5]. In this case, the reactants are mixed, pressed into a pellet and placed in a silica reaction tube which is cooled in liquid nitrogen before sealing to prevent sublimation of $ReCl_5$. The tube is then placed in a vertical tube furnace. The temperature is raised at 1°C min^{-1} to 850°C and kept there for 24 h before cooling at 6°C min^{-1}. With the furnace turned to almost horizontal, a temperature gradient (300–25°C) is used to remove volatile materials to the upper end of the tube to leave the product as a red–brown crystalline powder.

10.4.2 Metal containers

As mentioned above, high-temperature solid-state synthesis involving the early transition metals requires the use of metal containers, as exemplified in the extensive work on zirconium halide materials from which centred hexanuclear clusters can be excised [6]. Read the detailed description of procedures in reference 2 if you intend to carry out reactions in tantalum (or other refractory metal). In general, tubes are first cleaned with a solution consisting of 50% (by volume) conc. sulphuric acid, 25% concentrated nitric acid and 25% hydrofluoric acid (48% strength). (**Care!** HF causes severe burns. See Appendix A.) The bottom of the tube is then either crimped (Figure 10.3a) or capped (Figure 10.3b) and arc welded in helium or argon. After loading the tubes in a dry box, they must be sealed by arc welding unless the components of the reaction mixture have

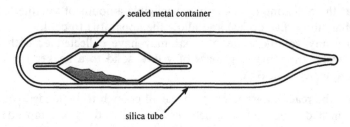

Figure 10.4 Arrangement for solid-state reactions in sealed metal tubes.

sufficiently low vapour pressures that open containers, perhaps blanketed by a noble gas, will suffice. Tubes are transferred to the welding apparatus from the dry box in a closed container after crimping the open end or fitting the end-cap. Any atmosphere from the box is removed by evacuation of the welder and back-filling with noble gas. The style shown in Figure 10.3b is more suitable for larger containers or higher pressures, while the evacuable design (Figure 10.3c) may improve transport rates. The welded metal container is protected from attack by the atmosphere at elevated temperatures by placing it inside a sealed silica or stainless steel jacket (Figure 10.4). After the reaction (which may take weeks), tubes can be opened with a spin-cutter of the type used for copper pipe.

References

1. Corbett, J.D. (1987) In *Solid State Chemistry: Techniques*, eds A.K. Cheetham and P. Day, Oxford University Press, Oxford, pp. 1–38.
2. Corbett, J.D. (1983) *Inorg. Synth.*, **22**, 15.
3. Nannelli, P. and Block, B.P. (1970) *Inorg. Synth.*, **12**, 170.
4. Zhang, X. and McCarley, R.E. (1995) *Inorg. Chem.*, **34**, 2678.
5. Gabriel, J.-C., Boubekeuer, K. and Batail, P. (1993) *Inorg. Chem.*, **32**, 2894.
6. Rogel, F., Zhang, J., Payne, M.W. and Corbett, J.D. (1990) *Adv. Chem. Ser.*, **226**, 369; Corbett, J.D. (1996) *J. Chem. Soc., Dalton Trans.*, 575.

11 Product Characterization

11.1 Introduction

Once the products of a reaction have been separated and purified, the next challenge is their characterization. Even compounds prepared by literature procedures must have their identity and purity verified. At this point a sound working knowledge of analytical techniques, encompassing both theory and practice, is essential. In addition, as a synthetic chemist it is vitally important that you are able to interpret the results from these analytical techniques and are aware of any inherent limitations to their use. This chapter does not include an in-depth discussion of theory; it is more than adequately covered elsewhere and I have assumed that readers will have had some theoretical background to common analytical methods (references are included for those who wish to refresh their memories). Details of interpretation have also been omitted. A broad coverage of most techniques from an inorganic viewpoint is given in the text by Ebsworth *et al.* [1], which also provides references to more detailed coverage by other authors.

The aim of this chapter is to establish an awareness of the various analytical techniques and provide a guide to the practical aspects of this crucial facet of research, i.e. how to prepare samples for analysis. It is important to obtain a sufficient set of data for each compound you prepare, and for full characterization (the determination of composition and structure), you will need to apply several techniques, some of which are more readily available than others. New compounds necessarily require a more extensive set of data than known compounds that have previously been fully characterized (although it is always worth checking that such compounds have been fully and correctly characterized – errors do occur). In general, you should obtain as much information as possible on your products. Analyse new compounds without delay; results from partially decomposed samples can cause a great deal of unnecessary confusion, and it is frustrating to have to prepare a fresh sample (usually when you can least afford the time) simply because one vital spectrum is missing!

11.2 Spectroscopic techniques

Advances in instrumentation and electronics over recent years have meant that most research departments have access to a range of sophisticated spectrometers and nowadays the first stage in the characterization of an organometallic or metalorganic compound is usually the measurement of ^{1}H-NMR and IR spectra. The presence or absence of expected functionalities and the relative amounts of any organic groups in the sample will normally enable you to decide whether further purification is necessary before elemental microanalysis and other physical measurements are carried out. The characteristic spectroscopic properties of groups or compounds can be found in several publications [2], and the introductory text by Williams and Fleming [2a] is a useful starting point when checking the NMR, IR, UV and mass spectroscopic properties of organic derivatives.

11.2.1   NMR spectroscopy

In the modern synthetic laboratory it is difficult to contemplate life without the NMR spectrometer. Of all the spectroscopic techniques, Fourier Transform NMR has developed most dramatically as a result of advances in technology [3], especially with the availability of more powerful computers and superconducting magnets. Whereas organic chemists restrict themselves in the main to ^{1}H- and ^{13}C-NMR, inorganic chemists use a much wider range of magnetic isotopes in the characterization of their compounds. Table 11.1 gives some idea of the possibilities.

Departmental arrangements for spectrometer access vary; in some departments researchers record their own spectra, whereas in others a service is provided. The norm is probably somewhere between these two extremes, but I would strongly recommend that you arrange to have some 'hands-on' training. The experience is invaluable. Possibly because the measurement of NMR spectra is now regarded as routine, research students often pay insufficient attention to sample preparation. With today's high-field instruments, sloppy technique at this stage can result in poor quality spectra, so to do justice to your painstaking synthetic efforts (after all, you may have spent weeks or months preparing the compound) it is worth putting some thought and care into the choice of a suitable solvent and the preparation of the sample. If you don't record your own spectra, you will probably find that the spectrometer operators will appreciate this extra effort on your part.

Choosing a solvent. Deuterated solvents are necessary for ^{1}H-NMR measurements so that the solvent peaks do not swamp those of the sample. In addition, a deuterium resonance provides a convenient means of locking the field/frequency, so some deuterated solvent is usually added when observing nuclei other than ^{1}H. The quantity needed for a stable lock will depend on the amount of deuterium in the compound, for

Table 11.1 Properties of useful NMR nuclei[a]

Isotope	Spin	Natural abundance N/%	Magnetogyric ratio $\gamma/10^7$ rad T^{-1} s^{-1}	Quadrupole moment $Q/10^{-28}m^{-2}$	Relative NMR frequency Ξ/MHz	Receptivity relative to ^{13}C D^c
^{1}H	1/2	99.985	26.7510	–	100.0	5680
^{2}H	1	0.015	4.1064	0.00273	15.4	0.00821
^{6}Li	1	7.42	3.9366	–0.0008	14.7	3.58
^{7}Li	3/2	92.58	10.396	–0.045	38.9	1540
^{10}B	3	19.58	2.8748	0.074	10.7	22.1
^{11}B	3/2	80.4	8.5827	0.0355	32.1	754
^{13}C	1/2	1.11	6.7263	–	25.1	1.00
^{14}N	1	99.63	1.9324	0.016	7.2	5.69
^{15}N	1/2	0.37	–2.7107	–	10.1	0.0219
^{17}O	5/2	0.037	–3.6266	–0.026	13.6	0.0611
^{19}F	1/2	100.0	25.1665	–	94.1	4730
^{23}Na	3/2	100.0	7.0760	0.12	26.5	525
^{27}Al	5/2	100.0	6.9706	0.149	26.1	1170
^{29}Si	1/2	4.7	–5.3141	–	19.9	2.09
^{31}P	1/2	100.0	10.829	–	40.5	377
^{51}Y	7/2	99.76	7.032	0.3	26.3	2160
^{53}Cr	3/2	9.55	–1.5120	0.03	5.7	0.49
^{55}Mn	5/2	100.0	6.598	0.55	24.7	994
^{57}Fe	1/2	2.19	0.8644	–	3.2	0.00419
^{59}Co	7/2	100.0	6.3171	0.40	23.6	1570
^{75}As	3/2	100.0	4.5816	0.3	17.1	143
^{77}Se	1/2	7.58	5.101	–	19.1	2.98
^{89}Y	1/2	100.0	–1.3106	–	4.9	0.668
^{93}Nb	9/2	100.0	6.5387	–0.2	24.4	2740
^{95}Mo	5/2	15.72	1.743	0.12	6.5	2.88
^{101}Ru	5/2	17.1	1.383	0.44	5.2	0.86
^{103}Rh	1/2	100.0	–0.8420	–	3.2	0.177
^{109}Ag	1/2	48.18	–1.2449	–	4.7	0.276
^{113}Cd	1/2	12.26	–5.9330	–	22.2	7.59
^{119}Sn	1/2	8.58	–9.9707	–	37.3	25.2
^{125}Te	1/2	6.99	–8.453	–	31.5	12.5
^{129}Xe	1/2	26.44	–7.3995	–	27.7	31.8
^{133}Cs	7/2	100.0	3.5089	–0.003	13.1	269
^{183}W	1/2	14.4	1.1131	–	4.2	0.0589
^{187}Os	1/2	1.64	0.6161	–	2.3	0.00114
^{195}Pt	1/2	33.8	5.7505	–	21.4	19.1
^{199}Hg	1/2	16.84	4.7690	–	17.9	5.42
^{205}Tl	1/2	70.50	15.438	–	57.6	769
^{207}Pb	1/2	22.6	5.5968	–	20.9	11.8

[a]Lanthanides have been omitted. Data from Harris, R.K, and Mann, B.E. (eds) (1978) *NMR and the Periodic Table*, Academic Press, London.

example less than 10% of C_6D_6 will suffice whereas a larger proportion of $CDCl_3$ is usually required. The solubility of your sample in deuterated solvents can usually (but not always) be deduced from the way it behaves in the non-deuterated analogues and a familiarity with the dielectric properties of the solvents (Appendix C) is useful in this regard. Apart from the requirement that your sample must dissolve, other factors also need to be considered when choosing an NMR solvent.

- Will the solvent react with the sample? Donor solvents may act as ligands, halogen abstraction can occur from $CDCl_3$ and alkoxide ligand exchange is likely to occur in methanol. In addition, exchangeable protons in the sample will not be observable in solvents that also contain exchangeable protons (e.g. water and methanol).
- Does the solvent contain the nucleus being observed? If so, you need to know the chemical shifts of the solvent resonances (Appendix D). This applies to the deuterated solvents used for ^{1}H-NMR, where molecules that are not fully deuterated give characteristic peaks in the spectrum. If the percentage deuteration is not sufficiently high, these peaks may obscure those of your sample.
- How pure is the solvent? Commercial suppliers give the percentage isotopic enrichment of solvents, but impurities such as water may also be present in unspecified amounts. Water can be particularly troublesome, not only for moisture-sensitive compounds but also for those containing exchangeable protons. It is therefore standard practice to dry NMR solvents, and for air-sensitive work they should also be degassed and used in an anhydrous environment in a dry box or on a Schlenk line. (Note that even if the sample is air-stable, removal of paramagnetic oxygen from the solvent is likely to be beneficial, although in some cases its presence may reduce the relaxation times of certain nuclei.) See section 6.2.2 for a discussion of drying agents, and vacuum-transfer from the desiccant is usually the most convenient option for the volumes of solvent involved; otherwise suspended solids must be removed by filtration before preparing the sample. Desiccants commonly used for NMR solvents include molecular sieves (although the dust from these can be difficult to remove), calcium hydride, activated alumina and Na/K. Solvents stored over a desiccant under nitrogen in small tap-tubes (Figure 11.1) are readily degassed by several freeze–pump–thaw cycles (section 5.3).
- How much does it cost? Deuterated solvents vary in price depending on the degree of deuteration and the difficulties associated with their synthesis. For example, chloroform-d (99.8 atom % D) is relatively cheap while tetrahydrofuran-d_8 (99.5 atom % D) is about one hundred times more expensive. Consequently, the expensive solvents should only be used when there are no other options; otherwise you may risk incurring the wrath of your supervisor.
- Is the liquid range and viscosity of the solvent appropriate for the NMR experiment being conducted? For variable temperature work, the melting and boiling points of the solvent must be outside the

Figure 11.1 Tap-tube for storing liquid reagents.

temperature range to be studied. The viscosity of the solvent can have a significant effect on linewidths, so where optimum resolution is important, non-viscous solvents such as acetone, acetonitrile, chloroform, dichloromethane or methanol should be used.

Another point to bear in mind is that the chemical shift of a particular resonance is not likely to be the same in all solvents. For example, ^{1}H-NMR spectra of compounds in aromatic solvents are often markedly different from those recorded in other solvents.

Choosing a tube. It is well worth browsing the catalogue of an NMR glassware supplier (e.g. Wilmad) to get an idea of the wide range of different NMR tubes that are available. The various types are grouped according to quality (i.e. the precision of their dimensions) and are supplied with thin, medium or heavy wall thicknesses. The most common outside diameters are 5 and 10 mm, but tubes with diameters up to 30 mm are also available for more specialist measurements. In general, measurements at higher fields will benefit more from the more expensive high quality tubes, but whether or not a tube will spin when inserted into the probe depends on the probe dimensions, and this can only be determined by experiment. A bad choice of tube will often result in a spectrum containing large spinning side-bands.

To obtain the best results from NMR tubes, they should be washed carefully to avoid scratching the glass, dried in a nitrogen stream or under vacuum and then kept free from dust (this also applies to new tubes taken straight out of the packet). The glass apparatus shown in

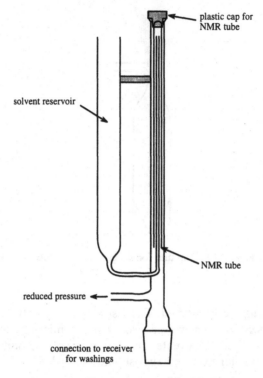

plastic cap for
NMR tube

solvent reservoir

NMR tube

reduced pressure

connection to receiver
for washings

Figure 11.2   Cleaning apparatus for NMR tubes.

Figure 11.2 provides a means of cleaning NMR tubes with a jet of solvent
and its use is described below.

- Push a plastic cap over the closed end of the NMR tube and then lower
  it over the inner jet tube so that it is held in place by the cap.
- Connect a water pump to the receiver flask and turn it on.
- Add solvent to the reservoir from a wash bottle. It will be sprayed into
  the NMR tube and fall into the reservoir.
- Dry the tube by pulling air through the apparatus for a while.

An ultrasonic cleaning bath is useful for the removal of more stubborn
residues from inside NMR tubes.

Choose a tube with a diameter to match the diameter of the NMR
probe being used. If an increase in pressure is likely during the measurement, use a medium- or heavy-walled tube. Having decided on the
diameter and wall thickness, the most important factor is how to seal the
tube. Push-fit polypropylene caps are usually supplied with new NMR
tubes, but these are of limited use for air-sensitive samples or long-term

storage, and their removal from thin-walled tubes often results in damage to the top of the tube. For samples that are fairly stable we use the soft natural rubber septa available from Aldrich. For air-sensitive compounds, a number of commercial designs have evolved to cater for different users, but we have found the screw-cap tubes from Wilmad (Figure 11.3a) to be very convenient and simple to use. The caps are available either with or without a hole for use with self-sealing septa. Another popular design for inert-atmosphere work incorporates a mini-Young's valve that can be connected to a manifold or vacuum line (Figure 11.3b), although these are comparatively heavy and we have experienced problems using them in an autosampler. A cheaper alternative is to flame-seal the NMR tube with a gas torch. If done properly, this ensures the total exclusion of air from a sample and is ideal for following reactions, variable-temperature measurements and long-term storage. When sealing samples in this way it is best to use medium-walled tubes, and to prolong their lifetimes we usually have our glassblower attach a 10 cm extension (this must be concentric and the glass must match that of the NMR tube). After adding the sample, the tube is sealed part-way up the extension (see below and Figure 11.7 for details) and after use the remainder of the extension is cut off and replaced, thereby minimizing the reduction in length of the precision glass tube.

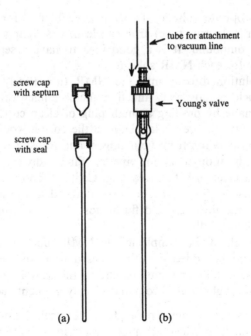

Figure 11.3   Re-sealable NMR tubes.

Sample preparation. The amount of material required for an NMR measurement depends on the sensitivity of the nucleus being observed (Table 11.1) and the type of NMR experiment being undertaken. For ^{1}H spectra on a high field instrument, as little as 1 mg is sufficient, but if the same sample is to be used for ^{13}C measurements, it is best to use a much more concentrated solution to reduce the number of scans and therefore the time needed to obtain the ^{13}C spectrum. You should also take into account the amount of the observed nucleus that is present in the sample, as the effective solution concentration is lowered significantly when only a few atoms are present in a high molecular weight compound. It is worth bearing in mind the following relationships (where S is the signal level and S/N is the signal-to-noise ratio):

- S/N $\propto \sqrt{}$(number of scans);
- S $\propto$ concentration of sample;
- number of scans $\propto$ time.

In simple terms, this means that doubling the concentration will enable the same signal-to-noise ratio to be obtained in a quarter of the time.

The following procedure is suitable for making up an NMR sample of an air-stable compound.

- Transfer the required amount of compound to a clean, small sample tube.
- Add the appropriate volume of solvent (*ca.* 0.5 cm^3 for a 5 mm tube, *ca.* 3 cm^3 for a 10 mm tube) from a clean, dry syringe and shake to dissolve the compound. It is a good idea to have a separate, labelled 1 cm^3 syringe for each NMR solvent.
- Filter the solution directly into the NMR tube (finely divided solids destroy the field homogeneity and thus increase linewidths). A convenient filter is made by pushing a small plug of clean cotton wool into a dropping pipette (Figure 11.4a). Handle the cotton wool with tweezers to avoid contamination from your fingers and transfer the solution into this filter with another clean pipette. Finely divided solid can be removed by a layer of a filter-aid (e.g. Celite) supported on the cotton wool, and pressure from a rubber teat attached to the top of the filter will speed up the filtration and flush most of the solution through the filter pad (Figure 11.4b).
- Check the depth of the sample in the NMR tube. If more solvent is required, it can be added directly or used to rinse the filter pad. In either case, after adding more solvent to the sample, ensure that the solution is thoroughly mixed to avoid density gradients within the tube.

The optimum sample volume depends on the probe being used, and the minimum volume to produce adequate resolution will be determined by experiment. For reproducible results, it is advisable always to use the

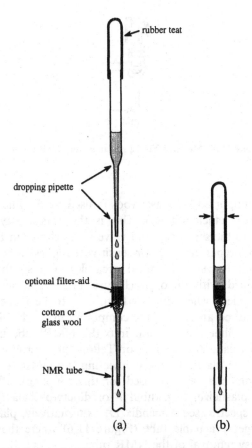

rubber teat

dropping pipette

optional filter-aid

cotton or
glass wool

NMR tube

(a)   (b)

Figure 11.4 Small-scale filtration with dropping pipettes.

same depth of solution for all measurements on a particular probe. When using minimum sample volumes in wider tubes, a PTFE plug with fins to hold it firmly in place is often inserted and pushed down to the meniscus to prevent the formation of a vortex during spinning. PTFE has a high coefficient of thermal expansion and this type of vortex plug should not be used for variable temperature work since at high temperatures the tube may crack, while at low temperatures the plug will fall to the bottom of the tube. If the plug is designed to be removed with a threaded rod, insert it at least part of the way with the rod attached. Since only one end of this type of plug is threaded, they can be difficult to remove if inserted upside-down.

The procedure described above can also be used in a dry box to prepare NMR samples of air-sensitive compounds, except that dried,

5 mm

Figure 11.5 Modified Young's adapter for NMR tubes.

degassed Celite supported on glass wool is used in the filter (glass wool alone is not a very efficient filter). This is the easiest way to add air-sensitive samples to the screw-cap and valve tubes shown in Figure 11.3.

For bench-top preparation of air-sensitive samples, attach an NMR tube to the manifold *via* a modified greaseless tap so that it can be evacuated and filled with nitrogen (Figure 11.5). This arrangement is particularly convenient when the NMR tube is to be flame-sealed. Add solvent to the required amount of the compound in a Schlenk flask under nitrogen and then filter the solution into the NMR tube using a small filter stick and a clean, dry cannula or Teflon tube. Ensure that solvent does not contact the plastic tubing connecting the NMR tube to the tap (a compression joint may also be used for this coupling), otherwise resonances from the plasticizer (e.g. dibutyl- or dioctylphthalate) will appear in the ^{1}H or ^{13}C spectra (see Appendix D). Alternatively, place the NMR tube inside a larger evacuable tube (Figure 11.6) where the lower screw cap allows for easy removal of the NMR tube.

When spectral resolution is not of paramount importance, it is possible to transfer good quality crystals directly to the NMR tube in the dry box and, if the compound is not very air-sensitive this can be done at the bench under a nitrogen purge as described below.

- Remove the cap from the flask containing the solid under a nitrogen purge and carefully remove the NMR tube from the tap adapter or threaded tube (Figures 11.5 and 11.6), briefly covering the open end with a finger while you insert this end of the tube into the neck of the flask.
- Holding the bottom of the NMR tube, select a few crystals and angle the tube so that they can be moved to the bottom of the tube with gentle tapping.
- Withdraw the tube from the flask, again covering the end with a finger while you reconnect it to the tap or replace it in the threaded tube.
- Evacuate the NMR tube and purge with nitrogen before unscrewing the tap and adding the measured amount of solvent from a syringe.

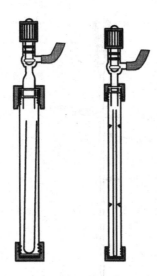

Figure 11.6 Apparatus for loading NMR tubes under an inert atmosphere.

With the correct depth of solution in the tube, carefully remove the tube under a nitrogen purge, hold the appropriate cap in the nitrogen flow then push it onto the open end of the tube. If the NMR tube is to be flame-sealed under reduced pressure, follow the procedure below, which is illustrated in Figure 11.7.

- With the tube still connected to the tap adapter and under nitrogen, cool the sample gradually from the bottom in liquid nitrogen.
- When the solution has solidified, evacuate the tube and close the tap.
- Set up a gas torch to give a small, hot flame, and carefully heat one side of the extension tube until you can see the glass being drawn inwards. Only leave the flame on the glass for a second or so at a time, otherwise there is a risk of the glass being drawn too far inwards or even being punctured.
- Repeat this process on the opposite side of the tube to create another indent facing the initial one.
- In the same manner, create another two indentations in the tube at 90° to the first two.
- Enlarge the gas flame slightly and heat the whole area around the four indentations while pulling the NMR tube gently downwards.
- As the glass softens and a neck forms, twist the tube and detach it, keeping the flame on the end of the tube to ensure an even airtight seal.
- Remove the sample from the liquid nitrogen and thaw it gradually from the top downwards (with care this can be done with fingertips).

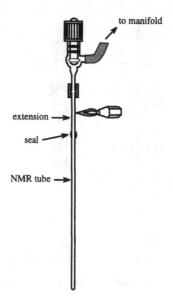

Figure 11.7   Flame-sealing an extended NMR tube.

This helps to prevent any expansion of the frozen solution from breaking the tube.

To open a flame-sealed NMR tube, score with a glass knife at or above where the extension was joined to the precision tubing, cool to reduce the internal pressure and then apply pressure as if to bend the tube so that it breaks at the score. Wrap the scored area in paraffin film to reduce the risk of damage to fingers or to dry-box gloves from broken glass.

In situ NMR studies.   The techniques described above are also used when following a reaction *in situ* by NMR. Solid reactants can be mixed in the tube before addition of the solvent, while a liquid reagent can be added from a syringe through a septum cap or in the dry box before the cap is fitted. Volatile reagents can be added from a calibrated line (section 5.3) to an NMR tube fitted with a Young's valve or a tap adapter before closing the valve or sealing off the tube as described above. Volatile products can be recovered from these tubes by vacuum transfer after connection to a vacuum line. Apparatus has been described [4] which enables flame-sealed tubes to be opened in an inert atmosphere after scoring with a glass knife, and a slightly modified version is shown in Figure 11.8. After connection to the vacuum line the apparatus is evacuated and the sample cooled in liquid nitrogen. The score on the NMR tube is aligned with the top of the inner glass tube to ensure a

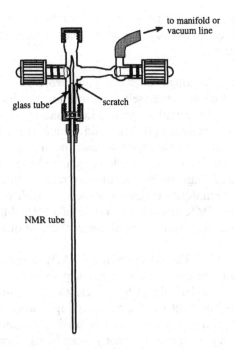

Figure 11.8 Opening a flame-sealed NMR tube under an inert atmosphere.

clean break when the tap is screwed in. On warming the NMR tube, volatiles can then be transferred to another flask for analysis (collection of non-condensable gases requires the use of a Töpler pump or adsorption trap).

Choosing a reference. Chemical shifts are normally reported relative to a standard reference, e.g. tetramethylsilane (TMS) for ^1H and ^{13}C, with positive values at higher frequencies than the reference peak. When reporting results, it is important to specify the reference for each spectrum. However, because of physical or chemical incompatibilities, it is not always possible to include the primary reference in the sample, and a compound with a known chemical shift relative to the primary reference must be used. Solvent resonances are frequently used as secondary reference peaks, and the protio impurities in deuterated solvents are particularly convenient for ^1H-NMR spectra (note that these peaks may be multiplets due to coupling to deuterium), e.g. $CHCl_3$ (singlet) in $CDCl_3$ or CHD_2CN (quintet) in CD_3CN. Spectra can also be recorded with the reference compound in a sealed, concentric capillary inside the NMR tube. For nuclei other than ^1H or ^{13}C, the 'sample replacement' method is often used. In this case the spectrometer is calibrated with a solution of the standard reference compound immediately before or after measurements on the sample.

Solid-state NMR. Although most of your work is likely to concern molecular species that lend themselves to solution studies, NMR spectroscopy can also provide useful structural information on solids, including polymers and oxide materials [5]. The solid sample is packed into a rotor to enable rotation at 5–7 kHz about an axis inclined at an angle of 54.7° to the magnetic field. This 'magic-angle' spinning (MAS) minimizes effects due to chemical shift anisotropy and thereby reduces linewidths. The use of this and other techniques to improve spectral quality is now enabling detailed solid-state NMR studies of a range of nuclei. Much information can be gleaned from such studies, particularly those of quadrupolar nuclei.

Paramagnetic samples. The presence in an NMR sample of a compound with unpaired electrons can cause variable degrees of line broadening, the magnitude of which is determined by the electronic spin-lattice relaxation time and the hyperfine electron-nuclear coupling. These effects on relaxation times and linewidths can sometimes be used to advantage. The relaxation times for ^{13}C nuclei in metal carbonyls can be significantly reduced by addition of $[Cr(acac)_3]$, resulting in improved S/N ratios, and in aqueous ^{17}O-NMR studies of oxo complexes, the signal due to free (or rapidly exchanging) water can be suppressed by addition of a soluble Mn(II) salt.

On those occasions when a spectrum of a paramagnetic compound can be obtained, the chemical shift range is generally greatly expanded (for example to hundreds of ppm for ^1H spectra). Applications of this effect include the addition of small amounts of europium or praseodymium complexes to samples with complicated, overlapping ^1H-NMR spectra to spread them out and aid in peak assignments, and the measurement of magnetic susceptibilities by NMR. The latter technique, developed by Evans [6], involves monitoring a particular peak in the ^1H-NMR spectrum of a compound and measuring the magnitude of the shift (Δv) caused by the presence of the paramagnetic material. Two solutions of an inert compound of equal concentration (often 2% *tert*-butanol) are prepared and a weighed amount of the paramagnetic sample is added to one of them. The ^1H-NMR spectra of both solutions are then recorded simultaneously by placing one in a capillary inside the other. The mass susceptibility χ of the paramagnetic substance is then given by:

$$\chi = \frac{3\Delta v}{2\pi v m} + \chi_0 + \frac{\chi(d_0 - d_s)}{m}$$

where m is the mass of material per cm^3 of solution, χ_0 is the mass susceptibility of the solvent and d_0 and d_s are the densities of the solvent and solution, respectively. The last term can be neglected for strongly paramagnetic substances such as transition metal complexes.

11.2.2 Vibrational spectroscopy

By using a combination of infrared (IR) and Raman methods, complementary information about the vibrational modes of a compound can be obtained, but by far the most common application of vibrational spectroscopy is the qualitative analysis of IR spectra, and compilations of characteristic group frequencies are available [2].

IR spectroscopy. IR spectrometers are readily available and until recently were mainly of the dispersive scanning type, in which a prism or grating separates the different frequencies in the mid-IR region (from 4000 down to 200 cm^{-1}). In contrast, most modern instruments are interferometers operating in the Fourier transform mode, making for greater sensitivity, flexibility, and ease of data manipulation. However, the majority of these instruments have a lower wavelength limit of 400 cm^{-1} and it becomes significantly more expensive to extend the range down to 200 cm^{-1}. Consequently, the older scanning instruments are still useful when bands in the region 400–200 cm^{-1} (e.g. metal-halogen stretches) are of interest.

Choice of window material. The most common window materials are NaCl, KBr or CsI, which can be used down to 600, 350 or 200 cm^{-1}, respectively. Single crystals of these materials with plane faces are used as windows in liquid cells or as support plates for liquids or mulls (see below). All are water-soluble materials, but CsI is the softest and by far the most expensive and need only be used when you are interested in the region from 350 to 200 cm^{-1}. In the absence of CsI plates or windows, it is possible to record a separate spectrum in this low frequency region using thin polyethylene plates. Salt plates require careful handling and should be stored in a desiccator. Always hold them by their edges, since moisture from fingers will damage the faces. When the plates show evidence of degradation or contamination they should be polished using a flat polishing block, fine abrasive powder and polishing cloths, taking great care to ensure that the surfaces of the plates are flat. IR spectra of aqueous solutions are measured using water-insoluble window materials such as CaF_2 and BaF_2.

Liquid samples. IR spectra of liquid materials are easily obtained (provided the sample does not react with the window material) by placing one or two drops onto a salt plate and placing another plate on top of this to form an even, thin film of liquid between the two plates, which are then carefully clamped together in a holder. The amount of material used should be such that none is squeezed out from between the plates when they are held together. Air-sensitive liquid films prepared in a dry box will usually survive in the air for long enough to record the spectrum, and afterwards it may be useful to separate the plates, expose the sample to air and record a second spectrum in order to get some idea of the degree of sensitivity.

Solution samples. Solids are usually prepared for IR analysis as a solution or as a suspension in an inert matrix. Solution spectra are often preferred, e.g. for metal carbonyl complexes since, although the bands are generally broader, they do not suffer from the splitting frequently observed in the spectra of crystalline solids. It is best to choose a solvent that does not absorb in the region of interest although with most FTIR spectrometers, a background spectrum of the cell filled with pure solvent can be subtracted from that of the cell containing the solution. Background spectra for a range of solvents can be recorded and stored in a computer for future use, assuming that the cell is properly cleaned on every occasion. Solvent absorptions are not so easily eliminated with a double-beam dispersive instrument since a cell identical to the sample cell must be available. This is then filled with pure solvent and placed in the second beam of the spectrometer.

Liquid cells comprise two windows held apart by a spacer, the thickness of which determines the cell volume (Figure 11.9). In order to achieve a leak-free seal without cracking the windows, these cells must be

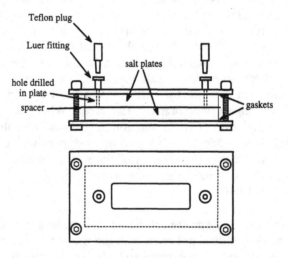

Figure 11.9 Liquid cell for IR spectroscopy.

assembled and tightened with great care. After choosing a suitable solvent, make up a solution of the solid, using inert atmosphere techniques if necessary, and transfer the solution into the clean liquid cell with a pipette or syringe. If a dry box is not available, an air-sensitive sample can be prepared at the bench as follows.

- Fit rubber septa over the Luer inlets of the cell.
- Insert a syringe needle attached to a hose from the manifold into one septum and a bleed needle into the other and purge the cell with nitrogen.
- Fill the cell from a syringe; insert the needle through one septum and allow an excess of solution to flow through the bleed needle into a small beaker to ensure the removal of all gas bubbles (don't let any solution run onto the outside of the cell).
- Remove the septa and cap the inlets with Teflon stoppers.

Mulls. It is usually more convenient to record the IR spectrum of a solid as a thin film of a mull, which is a finely divided suspension in an inert liquid support such as a hydrocarbon oil (Nujol) or hexachlorobutadiene (HCBD). Again, absorptions due to the matrix must be considered and in this case, they cannot easily be subtracted because of the difficulty in reproducing the film thickness. However, this is not usually a problem since the Nujol bands are readily identified (Figure 11.10) and if sample peaks of interest are obscured, a second spectrum can be recorded in HCBD (Figure 11.11).

A mull is prepared as follows, using a dry box for air-sensitive compounds.

- Grind 10–15 mg of the compound in an agate mortar to a fine powder (the smaller the particles, the better the spectrum).
- Put a drop of Nujol (or HCBD) onto the end of a microspatula and then transfer part of this onto the tip of the agate pestle and grind it into the solid (drops direct from a dropper tend to be too large, making it difficult to control the consistency of the mull).
- Repeat this process, adding small amounts of oil until the pestle leaves 'herring-bone' tracks in the mull. At this point the mull has about the correct consistency, although a slightly higher dilution should not seriously affect spectrum quality.
- Ensure that all the solid has been ground into the mull then scrape it off the end of the pestle and into the centre of the mortar with a microspatula.
- With the microspatula, scrape two or three drops of the mull from the mortar onto a KBr or CsI plate (being careful not to scratch the plate), add the top plate, rotating it to give a uniform thin film and then clamp the two plates in a holder.

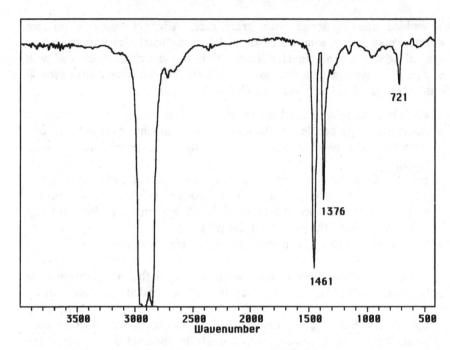

Figure 11.10 IR spectrum of Nujol.

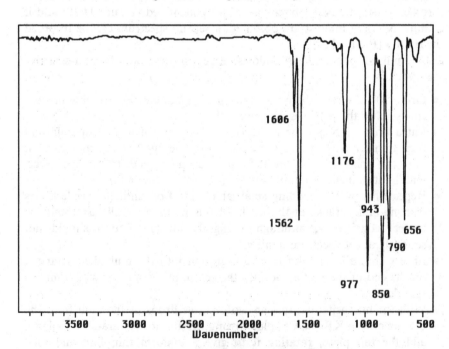

Figure 11.11 IR spectrum of hexachlorobutadiene.

Diffusion of air into the sample is usually sufficiently slow to allow plenty of time to record the IR spectrum before any decomposition occurs, and a spectrum of the decomposition products can be obtained subsequently by exposing the sample to air and then re-assembling the plates.

KBr discs. A matrix of KBr can also be used to support a finely divided solid sample for IR studies in the form of a disc (pellet), prepared as indicated below. This method cannot be used for compounds that are likely to undergo halide exchange and it is generally not suitable for air-sensitive materials.

- Grind about 10 mg of the solid to a fine powder in an agate mortar.
- Add about 100–200 mg of anhydrous KBr powder and continue to grind the two solids to produce a homogeneous mixture (as in the preparation of a mull, scattering is reduced if the sample is very finely ground).
- Assemble the two body parts of the stainless steel evacuable die (Figure 11.12), insert one of the stainless steel discs with its polished face upwards, push it to the bottom of the chamber with the plunger and then withdraw the plunger.
- Add the powdered sample and distribute it evenly (this can be done with one end of the plunger).
- Insert the other stainless steel disc, polished face downwards, and push it down onto the sample.
- Insert the plunger, place the O-ring around it and apply vacuum to the die (this removes any air and helps prevent oxidation of the sample under pressure).
- Place the die in a hydraulic press, close the release valve and apply the maximum working pressure to the die *via* the plunger for a minute or so.
- Remove the vacuum connection, open the pressure release valve and remove the die from the press.
- Pull the two halves of the die body apart, invert the upper half containing the sample and use a small plastic cylinder to ease out the metal discs and the sample disc in the press. Care is required to avoid breaking the KBr disc.
- Transfer the KBr disc to a sample holder and place it in the spectrometer.

It is usual for the transmission to be lower for KBr discs than for other media, but if the spectrum is poor because of too much or too little sample or if the disc is too thick, a new disc can be prepared after grinding the disc to a powder and adjusting the mixture if necessary.

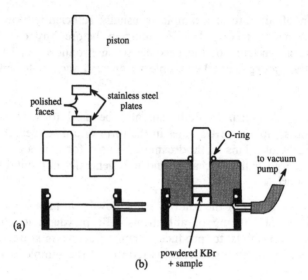

Figure 11.12 Die for making KBr discs for IR spectroscopy.

Gaseous samples. Occasions when you need to record the IR spectrum of a gas are likely to be rare, but to determine the identity of a gaseous reaction product, the sample must be loaded into a gas cell from the vacuum line. The basic cell consists of a high-vacuum tap attached to a glass tube which has salt windows sealed to each end with wax or O-rings and retaining plates (Figure 11.13). When wax seals are used, the windows are removed by warming with an IR lamp to avoid thermal stress which is likely to crack the salt plates.

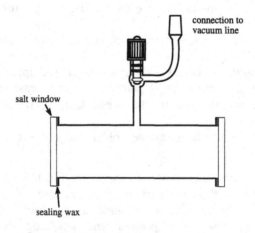

Figure 11.13 Gas sample cell.

Calibration of scanning instruments. It is good practice to include reference peaks in spectra recorded on scanning instruments. For each region of interest, choose a distinct peak in the spectrum of polystyrene (convenient peaks occur at 1603 and 1028 cm^{-1}) and, after recording the spectrum of the sample, raise the pen and rewind the chart paper to a few wave numbers before the chosen polystyrene peak. Remove the sample from the spectrometer, insert a polystyrene film and re-start the scan, stopping it immediately after the reference peak.

In situ IR studies. Specially designed cells allow IR spectra to be recorded under non-ambient conditions, i.e. high or low temperatures or pressures, and enable reactions to be monitored *in situ*, including the study of supported solid catalysts by diffuse reflectance. Whilst these techniques are beyond the scope of this book, experimental details have been published elsewhere and these should be consulted if you are likely to require such facilities [7,8].

Raman spectroscopy. Raman spectroscopy is not as widely used as IR spectroscopy and has not become a routine tool, so discussion here is limited mainly to some basic points. Raman spectrometers contain a laser to provide a single-frequency exciting radiation and a spectrophotometer to detect the Raman-shifted emission and, since the light used is in the visible region, it is possible to use glass optics, which is not the case in the IR region. Since glass containers pose no problems for Raman spectroscopy, liquid samples are easily handled and, given the fact that water has very weak Raman scattering, Raman has a significant advantage over IR for the study of aqueous solutions. Air-stable solid samples often do not need a container, otherwise they are sealed in glass capillaries. Coloured solids that strongly absorb the laser radiation will decompose due to local heating, although one way around this problem is to spin a disc of the sample in the beam to ensure that any one area is only irradiated for a short time. Fluorescent samples also cause problems because the fluorescence may swamp the much weaker Raman signal, whereupon the only cure is to change the wavelength of the exciting radiation.

11.2.3 Electronic spectroscopy

Most students are likely to have encountered UV/visible spectroscopy at undergraduate level but electronic spectra provide little direct information about molecular structure. The broad bands and complexity of interpretation make assignment difficult and this technique is more often used for following reactions once the spectrum of the starting material and/or product has been measured. To this end, cells for use with air-sensitive compounds have been described [8,9]. One example is shown in Figure

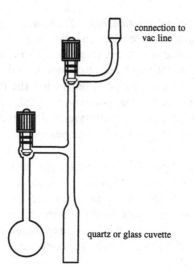

connection to
vac line

quartz or glass cuvette

Figure 11.14 Inert-atmosphere cell for UV–Vis spectroscopy.

11.14; the cell comprises a square glass or quartz cuvette for spectral measurements and a Pyrex bulb for separation of the solvent and addition of reagents and is used as described below.

- In a dry box, add the starting compound to the cuvette.
- Attach the cell to the high-vacuum line and add the degassed solvent by vacuum transfer (the volume can be measured by prior transfer into a calibrated tube).
- Record the UV/Vis spectrum of the starting compound.
- At the high-vacuum line, distil the solvent into the Pyrex bulb by cooling it in liquid nitrogen and then close the valve separating the bulb and cuvette.
- In the dry box, transfer the required amount of solid reagent to the cuvette (or alternatively add a liquid reagent to the solvent).
- Re-evacuate the cell on the vacuum line (after freezing the solvent if necessary), mix the contents by opening the lower valve and record the spectrum.

11.3 Mass spectrometry

The most frequent use of mass spectrometry in metalorganic chemistry is in the determination of molecular weights, although structural information can also be obtained by analysis of fragmentation patterns. A mass spectrometer records the mass-to-charge ratios (m/z) and relative abundances of ions obtained from a sample, and by far the most common method of ionization was, until recently, electron impact (EI). This involves the bombardment of a gaseous sample with a beam of electrons under high vacuum and the simplest ionization process (loss of an electron) will give a radical cation with essentially the same mass as the parent compound, i.e. the molecular ion. If the molecular ion decomposes it may not be present in the spectrum, and lower ionization energies must be used to minimize fragmentation. In EI, this is achieved by lowering the energy of the electron beam, but other 'softer' ionization techniques are now available which produce much more intense molecular ions. In chemical ionization (CI), for example, a beam of ions generated by electron impact is used to ionize the gaseous sample.

Samples for EI or CI must be volatile and may need to be heated, so these techniques are of limited use for the many metalorganic compounds that are involatile or thermally unstable. However, methods have been developed to enable ionization from the solid or liquid phase, and the range of energy sources used to desorb molecules or ions from the sample into the gas phase gives rise to a range of acronyms, e.g. field desorption (FD; application of a very high electric field), secondary ion mass spectrometry (SIMS; bombardment with a beam of ions), fast-atom bombardment (FAB; bombardment with a high-energy beam of neutral atoms of e.g. argon) and matrix-assisted laser desorption ionization (MALDI; laser ablation).

FAB [10] in particular has undergone extensive commercial development and is now available on most modern spectrometers. In this case, the sample is supported in a non-volatile matrix (usually glycerol or m-nitrobenzyl alcohol, NOBA) either as a mull (see section 11.1.2) or by dissolving the compound in a volatile solvent and adding this solution to the matrix. Either cations or anions of an ionic compound can be ejected from the surface and analysed, whereas neutral compounds undergo protonation or deprotonation to produce $[M+H]^+$ or $[M-H]^-$ ions. For EI and FAB ionization, it may be necessary to adapt the sample inlet so that air-sensitive samples can be loaded under an inert-gas purge.

Electrospray mass spectrometry (ESMS) [11] employs another soft ionization technique and was initially developed for the analysis of large polar biopolymers (molecular weights above 100 kDa have been measured) but promising results are now being obtained with coordination and organometallic compounds. ESMS is more convenient than FABMS in that solutions can be sampled directly under an inert gas, which eliminates problems due to matrix incompatibility and sample decomposition by e.g. hydrolysis. The sample is introduced into the spectrometer as a nebulized spray and the solvent is evaporated from the droplets as they are transmitted to the high vacuum section of the spectrometer. The ion beam is focussed by skimmer cones and the average kinetic energy is determined by the voltage applied to these cones. An advantage of the ES procedure is the limited amount of fragmentation, but this is influenced by the magnitude of the cone voltage, so it may be necessary to experiment with different values to obtain the best spectrum. The ability to sample directly from solution makes ESMS ideal for coupling to separation techniques, and a description of the on-line mass analysis of photochemical reaction products by ESMS demonstrates the further potential of this ionization method [12].

The interpretation of FAB and ES mass spectra is not always straightforward, since the charged species in the gas phase often consist of aggregated cations and anions. A feature of ESMS is the generation of multiply charged ions which appear in the spectrum as peaks with m/z values that are a fraction of the actual mass of the ion and with spacings in the isotope cluster pattern of 1/2, 1/3, 1/4, etc. For example, we have observed that the most intense cluster peak in the FABMS of $(Bu^n_4N)_2[Mo_6O_{19}]$ (positive-ion, NOBA matrix) is due to $\{(Bu^n_4N)_3[Mo_6O_{19}]\}^+$ and a smaller peak is due to $\{(Bu^n_4N)_2H[Mo_6O_{19}]\}^+$, while the negative-ion ES mass spectrum of an acetonitrile solution of this compound contains an intense cluster of peaks centred at m/z 439.8 due to $[Mo_6O_{19}]^{2-}$ (mass 879.6) [13] and a smaller cluster of peaks due to the ion pair $\{(Bu^n_4N)[Mo_6O_{19}]\}^-$.

11.4   Elemental microanalysis and molecular weight determinations

Many people will argue that until satisfactory elemental analyses have been obtained on a new compound, it cannot be considered to have been fully characterized. This is not a problem for stable, crystalline compounds but there are occasions where the physical and/or chemical properties of a compound make it difficult to obtain reliable micro-analytical results. You should therefore consider elemental analysis as an essential part of compound characterization unless it is impractical. Between 1 and 50 mg of material will be required for the analysis which will most likely be carried out by professional microanalysts within the department or (at greater expense) by an external commercial laboratory. It is, of course, possible to carry out your own analyses for certain elements, but much skill and expertise is required to work with small samples and achieve the necessary precision, so one-off analyses are likely to be time-consuming and use significant amounts of material. Brief descriptions of some of the techniques employed are given below, and if you intend to analyse your own samples, ref. 14 provides details of pro-cedures.

Before submitting a sample for analysis it must have been purified (section 8.3) and should be homogeneous. You are not likely to be held in very high regard by the analyst if your sample contains crystals of different colours, paper fibres or pieces of rubber septa. Sample tubes or ampoules should be scrupulously clean and dry, and if you load the sample in the dry box take great care not to introduce any impurities (static can be a problem in this regard).

11.4.1   CHN combustion analysis

The minimum microanalytical requirement for a metalorganic compound is the determination of the C, H and (if relevant) N content. Such analyses are carried out on a routine basis using an automated combustion analyser and a schematic representation of a Carlo Erba model is shown in Figure 11.15. About 1 mg of the pure compound is accurately weighed to 0.001 mg into a pure tin capsule which is then sealed by crimping. The capsule is introduced into the reaction furnace along with 10 cm^3 of oxygen whereupon flash combustion of the tin causes the local temperature around the sample to be raised to about 1600°C and the sample to be vaporized. Oxidation of the sample by a combination of the excess of oxygen and the solid oxidants in the packed column produces carbon dioxide, water, nitric oxide and nitrogen dioxide, while halogens are retained on the column. The helium carrier gas sweeps these gases into the reduction furnace where the nitrogen oxides are reduced to dinitrogen and in the final stage N_2, CO_2 and H_2O are separated on a GC column and detected quantitatively as they elute from the column. Several standards are loaded into the carousel with the samples for analysis to ensure that the machine is correctly calibrated.

Unless the analyser is attached to a dry box, tin capsules for the analysis of air-sensitive compounds are weighed on a microbalance before being transferred into the box where the required amount of approxi-

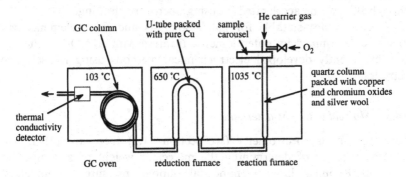

Figure 11.15 Schematic diagram of a CHN analyser.

mately 1 mg of each compound must be estimated (this improves with practise). The capsules are sealed with a special crimping tool, removed from the box and then re-weighed. During this process, the microbalance must not be disturbed or adjusted between the two weighings. Note that when using a nitrogen-filled dry box it is wise to run a blank (i.e. a capsule sealed in the dry box without any sample) to allow for the nitrogen content of the atmosphere above the sample.

11.4.2 Determination of elements other than C, H and N

If you have a requirement for the determination of elements other than C, H and N, the analysis will probably involve either atomic absorption or emission spectrometry. Both techniques involve the decomposition of the sample by intense heat into a gas containing atoms of the constituent elements. In atomic absorption spectrometry (AAS) the concentration of an element in this vapour is determined by measuring the absorption of light with a wavelength characteristic of that element. In atomic emission spectrometry (AES), higher temperatures cause excitation of the atoms which then emit radiation as they decay to lower energy states. By measuring the intensity of the characteristic emission lines it is possible to analyse for several elements concurrently. The high temperatures necessary for AES ($> 6000°C$) can be achieved in a plasma discharge, and in recent years inductively coupled plasma atomic emission spectro-scopy (ICP–AES) has become the method of choice for trace analysis in many analytical laboratories [15]. To take full advantage of this powerful technique some thought must be given to sample preparation. A solution of the sample with an accurately known concentration and containing about 10 ppm of the element of interest is required for injection into the plasma. The solvent used will depend on the standards available for that element and, since most of these are aqueous solutions, this may mean that the compound must first be decomposed by a suitable reagent (e.g. hot nitric acid) to render it soluble. Acid digestion may take some time, although this can be accelerated by using microwave heating.

ICP–AES is not suitable for halide analysis and these elements will usually be determined by potentiometric titration with $AgNO_3$ or alterna-tively by recently developed electrochemical methods using ion-selective electrodes.

11.4.3 Molecular weight determination

Apart from mass spectrometry (see section 11.3) there are other, less accurate, methods of determining the molecular weight of a compound based on Raoult's Law. Commercial vapour pressure 'osmometers' measure the temperature drop caused by evaporation of a solvent from a solution placed onto a thermistor. In operation, a solution of the sample of known concentration is compared with the pure solvent, and slightly air-sensitive compounds can be handled if the apparatus is purged with an inert gas (preferably argon, since it is denser than nitrogen). For soluble compounds, elevation of boiling point and depression of freezing point measurements can also be carried out on soluble compounds, but these methods are less than straightforward [16].

The Signer isopiestic (isothermal distillation) method relies on the diffusion of a solvent from one solution to another until the mole fraction of solute in each solution is the same. The apparatus shown in Figure 11.16 is loaded in the dry box and has two bulbs fitted with graduated capillary tubes (from pipettes) and a joint for connection to the vacuum line [16,17].

An accurately weighed amount of a standard compound such as ferrocene is placed in one bulb (0.010–0.015 g ± 0.1 mg) and the sample is similarly weighed out and placed in the other bulb. An inert solvent (0.5–0.8 cm^3) is placed in each bulb and the apparatus is assembled with the tap closed and removed from the dry box. The bulbs are cooled to freeze the liquid, and then evacuated on the line. The sealed apparatus is then allowed to warm to room temperature and set aside (preferably in the dark) for about a week to allow equilibration (although it may take longer). Volume changes are monitored periodically by tipping the solutions into the graduated limbs and when there is no further change the volumes are recorded. The molecular weight of the sample is then determined from the expression below.

$$MW_x = \frac{wt_x \cdot vol_s \cdot MW_s}{wt_s \cdot vol_x}$$

where: MW_x is the molecular weight of the compound, MW_s is the molecular weight of the standard, wt_x is the weight of the compound, wt_s

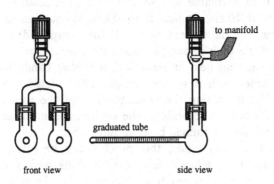

Figure 11.16 Signer apparatus for determination of molecular weights.

is the weight of the standard, vol_x is the volume of the solution of the compound and vol_s is the volume of the solution of the standard.

11.5 Growing crystals for single-crystal X-ray diffraction

Single-crystal X-ray diffraction is a very powerful technique for structural characterization and provides you with a picture of the molecule at the end of the process, but you must not fall into the trap of thinking that a crystal structure constitutes full characterization. It is always possible that the crystal chosen is not representative of the whole sample and is only a minor impurity that happens to crystallize nicely. You should, therefore, regard a crystal structure determination as just part of the characterization of a new compound.

As a preparative metalorganic chemist you are likely to spend a significant amount of your time in efforts to obtain good quality crystals of the appropriate size for X-ray crystallography. This is an art which some people master more easily than others and it can be a very frustrating business, but patience is usually rewarded. The fundamentals of crystallization have been covered in section 8.3.1 and suitable crystals may be obtained from the routine purification process, but the small-scale variations described here are useful when the crystallization must be controlled more carefully. A small hand lens with × 10 magnification is indispensable for examining crystals, and I recommend that you buy your own and label it clearly (everyone else will want to borrow it).

11.5.1 Slow evaporation

If the compound is air-stable and you have identified a suitable solvent pair, dissolve it in a volatile solvent and transfer a small amount of the solution to a 5 or 10 cm^3 conical flask. Then, without causing precipitation, add a less volatile solvent in which the compound is less soluble, seal the neck of the flask or sample tube with plastic film (this can be cut from a plastic bag and held in place with a rubber band) and pierce the plastic a few times with a syringe needle. The volatile component will evaporate slowly, increasing the concentration of the solute and at the same time decreasing its solubility. The optimum concentration to grow good crystals will be determined by trial and error.

For air-sensitive compounds, this process can be carried out under nitrogen by placing the solution in a small Schlenk flask and then establishing a slow flow of nitrogen from the line over the solution and out through an oil bubbler to the exhaust line.

11.5.2 Vapour diffusion

For small-scale crystallizations, we have found that the fairly simple arrangement for vapour diffusion shown in Figure 11.17a can be used for many air-sensitive compounds. The inner glass tube should be sufficiently long so as to be removed easily, and the diameter can be chosen to suit the amount of sample used. After placing the tube inside the flask, evacuate and fill the apparatus with nitrogen in the usual way. Make up a solution of the compound in a separate flask and transfer a small amount to the inner tube *via* syringe or cannula. Add a volatile solvent in which the compound is poorly soluble to the flask (around the outside of the tube), and seal the flask. With this particular arrangement, the volatile solvent will diffuse into the inner solution to make the solute less soluble and during this time the level of liquid in the inner tube will rise, so do not add too much solution initially. When the sample has crystallized, open the tap to nitrogen, remove the solvent from around the tube and then remove the mother liquor from the inner tube with a syringe or cannula (this may need to be left in place for solvent-dependent crystals). The adapter shown in Figure 11.17b allows for easier manipulation of the crystallized sample and can also be used for larger scale crystallizations (compare with Figure 8.3).

Crystals can be removed under nitrogen with a long, thin spatula or a pointed length of wire, either in the dry box or directly into a petri dish containing an inert, viscous oil (see section 11.5.3 below).

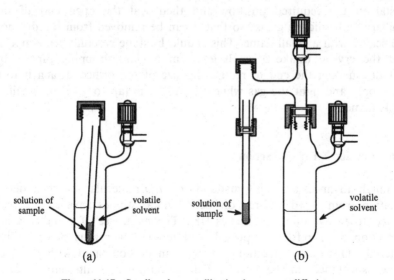

solution of sample   volatile solvent   solution of sample   volatile solvent

(a)   (b)

Figure 11.17 Small-scale crystallization by vapour diffusion.

The small diameter tube in this arrangement can also be used for small scale recrystallization of air-sensitive compounds by liquid layer diffusion (no solvent is added to the reservoir in this case). When setting up recrystallizations, it is advisable to clamp the flask at eye level so that you can easily monitor progress with the hand lens without disturbing the flask.

11.5.3 Mounting crystals

It is much easier to mount crystals for X-ray diffraction studies if the diffractometer is fitted with a gas-stream cooling system. In this case, crystals can be transferred directly into an inert, viscous oil (e.g. a perfluoropolyether) contained in a petri dish. A suitable crystal is selected under a microscope, transferred to the mounting pin and then quickly set up in the cooling stream so that the oil freezes.

If the diffractometer is not fitted with such a cooling system, the crystal must be sealed in a Lindemann capillary or in an epoxy resin. Transferring a crystal into a capillary is a tedious affair, especially in a dry box, and is a severe test of patience and temper. If you ever have to do this you will need several capillaries of different diameters and some lengths of glass fibre (drawn from a thin rod) which will fit inside the capillaries. To hold the crystal in place during data acquisition, smear a **very small** amount of silicone grease on one part of the inside wall of the capillary using a glass fibre. Select an appropriate crystal with a pointed wire, transfer it to a glass fibre (put a tiny smear of silicone oil on the end of the fibre) and then insert it into the capillary. By rotating the fibre, detach the crystal and withdraw the fibre. Use a clean fibre to push the crystal to the required position and then seal the open end of the capillary with silicone grease so that it can be removed from the dry box and sealed with a small flame. This should be done carefully so as not to heat the crystal (make a small jet from a glass dropping pipette by partially melting the end to reduce the size of the orifice, attach it to a gas supply and light the gas while adjusting the tap to give the smallest stable flame).

11.6 Powder X-ray diffraction

Although advances are being made in *ab initio* molecular structure determination from powder X-ray diffraction data, studies have so far been limited to fairly simple molecules [18]. The main use of powder X-ray diffraction is, therefore, to provide diffraction fingerprints of crystalline materials. Observed diffraction patterns can be compared with those of known substances in the Powder Diffraction File, or alternatively with patterns computed from proposed structures. Note that this provides a means of establishing the identity of insoluble precipitates formed in reactions where there might be some ambiguity (is it really lithium chloride?). Moreover, with the increasing emphasis on materials-related research, it is important for those involved to have a working knowledge of this basic solid-state characterization technique [19].

Powdered samples can be sprinkled onto double-sided adhesive tape or onto a glass slide smeared with grease. The thin layer of randomly oriented crystals is then placed in the X-ray beam. Thin films of materials deposited onto substrates by a variety of growth techniques can also be studied. Most modern instruments plot the intensity of the diffracted beam against the scattering angle (2θ) or the lattice plane separations (d), and any preferred orientation of the lattice (important for certain materials applications) will result in enhanced intensities of certain peaks. Components of mixtures can be identified from their characteristic patterns and if the diffractometer is fitted with a variable-temperature hot stage, the interconversion of different phases or compounds with temperature can be monitored. More complex analysis of the diffraction pattern involves fitting the observed peaks to those calculated for a given structure (Rietveld analysis) [20].

11.7 Other techniques

In this section, brief mention is made of various other characterization techniques which, although generally less frequently used than those already described, may find routine use in certain areas of research.

11.7.1 Conductivity measurements

There may be occasions when you need to confirm that a compound is ionic and/or determine the number of constituent ions without resorting to a crystal structure determination. Such a situation might arise, for example, when it is unclear whether an anion is bonded to the metal or simply present in the compound as a counterion, and this can often be resolved by conductivity measurements. A conductivity cell is shown in Figure 11.18. The resistance (R) of an electrolyte is measured using an A.C. bridge circuit, and is related to the resistivity ρ (defined as the resistance of the solution in a cell with 1 cm × 1 cm electrodes that are 1 cm apart) by the cell constant C.

$$R = C\rho$$

If C is not known it must first be determined by measuring the resistance of a standard electrolyte such as 0.02 M potassium chloride solution, which has a conductivity κ (which is $1/\rho$) of 2.768×10^{-3} S (the unit for

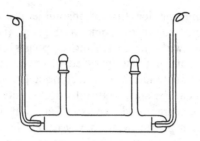

Figure 11.18 Conductivity cell.

conductivity is the siemens S; $1S = 1\Omega^{-1}$). The cell constant can then be calculated by using the expression:

$$C = 0.002768R$$

Hence the conductance κ of any solution in the cell can be determined, but this must be standardized to enable comparison with other substances and is therefore converted to molar conductance Λ, defined as the conductance of a 1 cm^3 cube of solution containing one mole (or formula weight) of solute. If the molar concentration of the electrolyte is c, then:

$$\Lambda = 1000C/c$$

However, an assumption of the formula weight may well be erroneous and a single measurement may lead to incorrect conclusions about the nature of the electrolyte. A more thorough treatment uses Kohlraush's law for strong electrolytes:

$$\Lambda = \Lambda_0 - Kc^{1/2}$$

Λ is determined for a range of concentrations in the region of 10^{-3} M and is then plotted against $c^{1/2}$. The intercept is Λ, the molar conductivity at infinite dilution, and the slope K depends upon the nature of the solvent and the electrolyte. The solution should be thermostatted at 25°C as Λ varies with temperature. This then allows the application of the Debye–Hückel–Onsager theory which is an attempt to quantify the effects of solvent and electrolyte properties on K in the expression above:

$$K = (A + \omega B\Lambda_0)$$

and hence

$$\Lambda_0 - \Lambda = (A + \omega B\Lambda_0)c^{1/2}$$

A plot of $(\Lambda_0 - \Lambda)$ against $c^{1/2}$ then gives a straight line with slope $(A + \omega B\Lambda_0)$. Assumptions about formula weights (and hence about electrolyte type) can then be checked against values of $(A + \omega B\Lambda_0)$ for known compounds containing similar counter-ions and in similar solvent systems. Measurements on a wide range of complexes in different solvents have been compiled [21] and this reference should be consulted if you intend to carry out conductivity measurements. Non-aqueous solvents of choice are nitromethane and acetonitrile but their toxicity must be taken into account when using them.

11.7.2 *ESR spectroscopy*

If your work involves compounds with unpaired electrons then ESR spectroscopy provides a means of investigating spin delocalization and the interactions between electrons and spin-active nuclei, and complements solid-state magnetic studies. In general, compounds that give useful ESR spectra will have extremely broad or unobservable resonances in their NMR spectra (see section 11.2.1). Measurements are normally carried out on powders or frozen solutions, although it is possible to obtain spectra of single crystals. Cells are constructed from silica, as borosilicate glass contains paramagnetic impurities, and samples must be degassed to remove any oxygen. A typical design is shown in Figure 11.19. Detailed discussion of this technique is beyond the scope of this book, but ref. 1 describes the basic principles and provides examples of applications with further references to more substantial texts on the subject.

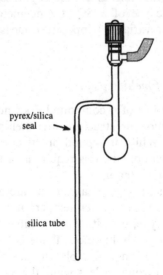

pyrex/silica
seal

silica tube

Figure 11.19 ESR cell.

11.7.3 Mössbauer spectroscopy

Although Mössbauer spectroscopy (γ-ray spectroscopy) is not commonly available, its most frequent use is in the study of solid compounds of iron (^{57}Fe) and tin (^{119}Sn) and provides information about the oxidation state and environment of the metal. Other isotopes that may be useful for Mössbauer studies include ^{99}Ru, ^{121}Sb, ^{125}Te, ^{129}I and ^{197}Au and more extensive lists of suitable isotopes are given in standard texts [22].

11.7.4 Thermogravimetry

Thermal analysis methods are discussed in more detail in section 12.3, but thermogravimetry, TG (commonly referred to as thermogravimetric analysis, TGA) is mentioned here because it provides a convenient means of establishing the amount of solvent of crystallization present in a compound. This is particularly useful for hydrated compounds or in cases where the amount of an organic solvent cannot be determined by NMR. Practical details are given in section 12.3.

11.7.5 Magnetic measurements

The Evans method for determining magnetic susceptibilities in solution by NMR is described in section 11.2.1. Simple measurements on solids can be made using the Gouy method, and commercial bench-top instruments are available (such as the Johnson–Matthey magnetic balance). However, this technique has its limitations and more sophisticated methods (e.g. the Faraday method or the use of a SQUID magnetometer) must be used when high sensitivity or variable temperature measurements are required [23].

11.7.6 Determination of melting point

Although the melting point of a compound does not give direct information about its composition or structure, the melting range may give an indication of its purity. Whilst it is good practice to obtain melting points for all air-stable new compounds, this requirement tends to be less rigidly enforced for air-sensitive materials.

Various designs of melting point apparatus are available which enable the sample to be observed and its temperature measured while it is heated. In the simplest type, a melting point tube that has been sealed at one end by heating in a partly luminous flame is loaded with the crushed sample to a depth of 5–10 mm and placed into a metal heating block. The sample is observed through a lens while the temperature is monitored during heating. Melting-point tubes can be loaded with air-sensitive compounds in the dry box, sealed with silicone grease and then sealed with a microflame upon removal from the box. A more sophisticated apparatus consists of a hot-stage and a microscope fitted with polarizing filters. In this case, the crushed sample is placed onto a microscope slide under a cover-slip. By observing the sample through the microscope with polarized light during heating, phase changes (such as those shown by liquid crystals) can be detected by changes in the appearance of the rather beautiful patterns which result, while at the melting point, the isotropic liquid will appear dark. In some cases, compounds can be seen to undergo reactions (one that I remember from my PhD studies is the cyclometalation of a coordinated phosphine) and the products may sublime onto the cover-slip. More sophisticated studies of this nature are carried out using thermal analysis techniques (section 12.3).

References

1. Ebsworth, E.A.V., Rankin, D.W.H. and Cradock, S. (1991) *Structural Methods in Inorganic Chemistry*, 2nd edn, Blackwell Scientific Publications, Oxford.
2. (a) Williams, D.H. and Fleming, I. (1995) *Spectroscopic Methods in Organic Chemistry*, 5th edn, McGraw Hill, Maidenhead; (b) Socrates, G. (1994) *IR Characteristic Group Frequencies*, 2nd edn, Wiley, Chichester.
3. Derome, A.E. (1989) *Modern NMR Techniques for Chemistry Research*, Pergamon Press, Oxford.
4. Bergman, R.G., Buchanan, J.M., McGhee, W.D., *et al.* (1987) In *Experimental Organometallic Chemistry: A Practicum in Synthesis and Characterization*, eds A.L. Wayda and M.Y. Darensbourg, ACS Symposium Series, Vol. 357, p. 227.
5. Harris, R.K. (1983) *Nuclear Magnetic Resonance Spectroscopy*, Pitman, London.
6. Evans, D.F. (1959) *J. Chem. Soc.*, 2003.
7. Darensbourg, D.J. and Gibson, G. (1987) In *Experimental Organometallic Chemistry: A Practicum in Synthesis and Characterization*, eds A.L. Wayda and M.Y. Darensbourg, ACS Symposium Series, Vol. 357, p. 230; Schenk, W.A., *ibid.*, p. 249.
8. References to special apparatus are also given in ref. 1 and in Shriver, D.F. and Drezdzon, M.A. (1986) *The Manipulation of Air-Sensitive Compounds*, 2nd edn, Wiley, New York.
9. Marshall, J.L., Hopkins, M.D. and Gray, H.B. (1987) In *Experimental Organometallic Chemistry: A Practicum in Synthesis and Characterization*, eds A.L. Wayda and M.Y. Darensbourg, ACS Symposium Series, Vol. 357, p. 254.
10. Miller, J.M. (1984) *Adv. Inorg. Chem. and Radiochem.*, **28**, 1.
11. Hofstadler, S.A., Bakhtiar, R. and Smith, R. (1996) *J. Chem. Educ.*, **73**, A83 (a good introduction to ESMS which includes leading references to this and other soft ionization techniques); Hop, C.E.C.A. and Bakhtiar, R. (1996) *J. Chem. Educ.*, **73**, A162 (examples of ESMS in inorganic and polymer chemistry).
12. Arakawa, R., Jian, L., Yoshimura, A., *et al* (1995) *Inorg. Chem.*, **34**, 3874.
13. Lau, T.-C., Wang, J., Guevremont, R. and Siu, K.W.M. (1995) *J. Chem. Soc., Chem. Commun.*, 877.
14. *Vogel's Textbook of Quantitative Chemical Analysis*, (1989) 5th edn, revised by G.H. Jeffery, J. Basset, J. Mendham and R.C. Denney, Longman, Harlow, 1989.
15. Boss, C.B. and Fredeen, K.J. (1989) *Concepts, Instrumentation and Techniques in Inductively Coupled Plasma Atomic Emission Spectrometry*, Perkin Elmer; Thompson, M. and Walsh, J.N. (1983) *A Handbook of Inductively Coupled Plasma Spectrometry*, Blackie, Glasgow.

16. Shriver, D.F. and Drezdzon, M.A. (1986) *The Manipulation of Air-Sensitive Compounds*, 2nd edn, Wiley, New York, p. 38.
17. Burger, B.J. and Bercaw, J.E. (1987) In *Experimental Organometallic Chemistry: A Practicum in Synthesis and Characterization*, eds A.L. Wayda and M.Y. Darensbourg, ACS Symposium Series, Vol. 357, p. 94.
18. Lightfoot, P., Glidewell, C. and Bruce, P.G. (1992) *J. Mater. Chem.*, **2**, 361.
19. West, A.R. (1984) *Solid State Chemistry and its Applications*, John Wiley and Sons, Chichester.
20. Rietveld, H.M. (1969) *J. Appl. Crystallogr.*, **2**, 65.
21. Geary, W.J. (1971) *Coord. Chem. Rev.*, **7**, 81.
22. Greenwood, N.N. and Gibb, T.C. (1971) *Mössbauer Spectroscopy*, Chapman and Hall, London; Bancroft, G.M. (1973) *Mössbauer Spectroscopy*, McGraw-Hill, London; Gibb, T.C. (1976) *Principles of Mössbauer Spectroscopy*, Chapman and Hall, London.
23. Cheetham, A.K. and Day, P. (eds) (1987) *Solid State Chemistry Techniques*, Oxford University Press.

12 Special Techniques

12.1 Introduction

The wide and varied nature of chemistry is such that undoubtedly there will be occasions when your work will require an unfamiliar technique. In such cases, someone else in the laboratory (perhaps a post-doctoral worker with expertise from another research group) may be able to advise and assist you; otherwise you will have to find and read the relevant literature to familiarize yourself with the technique, perhaps with a view to setting up a new facility. The aim of this chapter is therefore to provide brief overviews of selected techniques that are used more frequently in some laboratories (often on a routine basis) than in others. Even if you presently have no perceived need for any of these techniques, it is advisable to develop an awareness of innovative methods which, with widespread use, may someday become routine tools of the synthetic chemist. Electrochemical measurements are included here rather than in chapter 11 since they are more frequently used as investigative techniques, rather than for compound characterization.

12.2 Electrochemical techniques

If you intend to carry out electrochemistry on your compounds, I recommend that you first consult colleagues who already have experience with the techniques, and set aside some time to read some of the more detailed discussions of cell design and electrode selection [1–3]. Electrochemistry can be very frustrating for the novice who is not aware of the pitfalls.

12.2.1 Cyclic voltammetry

Cyclic voltammetry (CV) is the most common method for measuring the redox properties of a compound in solution, and the article by Mabbott is a useful starting point for first time users [4]. In a CV experiment, the current flowing between two electrodes is monitored while the applied potential is ramped between two pre-set values one or more times. The cell (an example is shown in Figure 12.1) is operated under an inert

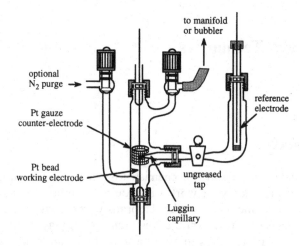

Figure 12.1   Electrochemical cell for cyclic voltammetry.

atmosphere to protect air-sensitive compounds and eliminate currents due to the reduction of O_2, and is fitted with a gas purge inlet. Cells can also be designed for use on a high vacuum line. The reference electrode should provide a stable potential. In non-aqueous solvents where anhydrous conditions are critical, a silver quasi-reference electrode (AgQRE) is often used. Formed by immersing a silver wire in an acetonitrile solution of the electrolyte, the potential of this electrode is not as highly reproducible as other electrodes, and it is usual to check the potential of a standard couple such as Cp_2Fe/Cp_2Fe^+ during the experiment. A more reliable Ag/Ag^+ reference electrode is prepared by immersing the silver wire in a solution of electrolyte that is also 0.001–0.002 M in $AgNO_3$. The ungreased ground-glass tap shown in Figure 12.1 is closed but wetted with a film of electrolyte solution. Together with the fine frit in the reference electrode, this helps reduce the diffusion of Ag^+ into the cell, which could cause problems at the working electrode.

The potential at the working electrode with respect to the reference electrode is controlled and monitored very precisely through the potentiostat, while a ramp voltage is provided by a waveform generator. The current flowing between the working electrode and the counter electrode is measured and plotted against the applied voltage to give a voltammogram. Most modern systems are interfaced with a computer, which simplifies parameter control and enables flexible data manipulation and storage prior to printing the voltammogram. Alternatively the output can be fed directly to a chart recorder.

To carry out CV measurements, a degassed solution of a supporting

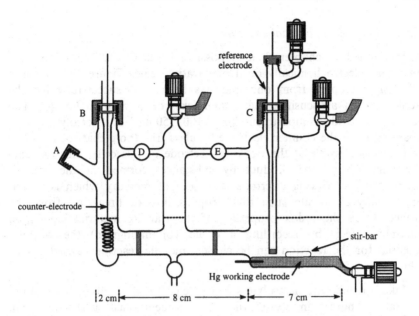

Figure 12.2 Electrochemical cell for controlled potential electrolysis.

electrolyte is prepared in a solvent with a moderate dielectric constant and a low proton availability. Of the wide range of solvents that find use in electrochemistry, acetonitrile is probably the most convenient (others that are widely used include propylene carbonate, dimethylformamide and dimethyl sulphoxide) and it should be purified before use (see section 6.2). 1,2-Difluorobenzene has also been recommended as a relatively inert, non-coordinating solvent for electrochemical studies on transition-metal complexes, although it is expensive and must be collected and distilled for re-use [5]. A common electrolyte is $Bu^n_4N^+BF_4^-$, which should be recrystallized from ethyl acetate/pentane. You should gain familiarity with the system by filling the cell with supporting electrolyte and recording several voltammograms. This will also reveal the presence of impurities from the electrolyte, the solvent or from a cell that has not been properly cleaned and dried. When you are happy that the system is operating correctly, add your sample to the solution with stirring until it is completely dissolved, turn off the stirrer and carry out the CV measurements. An *in situ* reference potential can be obtained by recording a voltammogram after adding ferrocene to the solution when all other measurements are complete. Measurements on air-sensitive compounds can also be carried out in a dry box fitted with suitable electrical breakthrough leads for connection to an external potentiostat.

12.2.2 Controlled potential electrolysis

As mentioned in chapter 6, it is possible to carry out oxidations and reductions electrochemically on a preparative scale. Figure 12.2 shows a cell design adapted from one that has been used successfully for the preparation of air-sensitive metal complexes on a vacuum line [6]. The working electrode compartment holds 100–150 cm^3 of electrolyte and is separated from the counter electrode by glass frits (note that reduction or oxidation is possible at the working electrode of this cell). This prevents contamination of the product by substances formed at the counter electrode. The working electrode is a pool of mercury which is stirred magnetically at a rate such that droplets do not break off from the surface. Once the redox properties of the starting material have been determined (e.g. by recording a cyclic voltammogram) the working potential for the reaction can be chosen and the reaction carried out as below.

- Make up sufficient electrolyte solution, e.g. 0.1 M tetrabutylammonium tetrafluoroborate in acetonitrile, in a Schlenk flask under nitrogen. Perchlorate salts should not be used for large-scale CPE.
- In a separate flask, make up a solution of the starting compound using some of the electrolyte solution.
- With the electrodes removed, add mercury to the working electrode compartment.
- Cap the screw joints **A**, **B** and **C**, connect the cell to the manifold and evacuate and fill with nitrogen in the normal fashion.
- Add degassed electrolyte solution to the compartments of the cell *via* cannula so that the liquid levels are equalized.
- Fit the reference and counter electrodes at **B** and **C** under a stream of nitrogen and make electrical connections to all three electrodes.
- Add the solution of starting compound to the working electrode compartment.
- Stir the surface of the mercury at a rate just below that required to cause drops of mercury to separate.
- Close the Teflon taps **D** and **E**.
- Start the electrolysis and monitor the current.
- Stop the electrolysis when the current has decayed to a steady state and drain the mercury from the cell.
- Transfer the solution in the working electrode compartment to a flask *via* cannula for work-up.

More specialized electrochemical studies are beyond the scope of this book, but the text by Christensen and Hamnett describes a variety of advanced experimental techniques that provide detailed information about redox processes occurring at the electrode [1].

12.3 Thermal analysis

As synthetic chemists become increasingly involved with new developments in materials chemistry, they must become familiar with the variety of methods used to measure physical properties. Thermal analysis techniques are used routinely to detect phase changes in liquid crystals and to study the conversion of molecular precursors to materials.

Thermal analysis is the collective term for a wide range of techniques used to monitor changes in physical properties as a sample is heated. In addition to the three most common measurements of mass (thermogravimetry, TG), temperature (differential thermal analysis, DTA) and enthalpy (differential scanning calorimetry, DSC), changes in the mechanical, optical, magnetic, electrical and acoustic properties of a substance can be monitored and gases evolved during heating can be analysed.

The increase in the thermal motion of constituent atoms, molecules or ions when a pure substance is heated may cause changes in the crystal structure or induce the substance to sinter, melt or sublime. In addition, decomposition may result in the formation of new molecular species and, if these are volatile, the sample will lose weight.

This section is limited to brief descriptions of TG, DTA and DSC and associated equipment, but for those who wish to know more about thermal analysis, two texts are available which provide very good overviews, ranging from basic theory to advanced applications [7,8].

12.3.1 Thermogravimetry

Measurements of mass changes during heating are made with a thermobalance, the main features of which are described in section 12.3.3. The output TG curve is a plot of either the actual mass or the percentage of the original mass against temperature and provides information about drying processes, loss of solvent of crystallization (and hence the amount of solvent present in the sample), thermal decomposition reactions (more information can be obtained with simultaneous evolved gas analysis, EGA) and oxidation reactions. The output from the balance can also be differentiated electronically to give a derivative thermogravimetric (DTG) curve, making it easier to interpret complex TG curves. Digital TG data can be converted to DTG curves within the software supplied with the instrument or other graphing packages.

12.3.2 Differential thermal analysis and differential scanning calorimetry

In DTA, the sample and an inert reference material (alumina is often used) are both subjected to the same heating programme and the difference in temperature ΔT between them is recorded. An endothermic event in the sample will cause the temperature of the sample to lag behind that of the reference material, and if ΔT is plotted against the reference temperature (or the furnace temperature) a deviation from a horizontal line will occur. An exotherm will produce a peak in the opposite direction. DTA curves should be marked with either the endo or exo direction. The area under a peak is related to the enthalpy change ΔH of the process.

In classical DTA, thermocouples are placed directly in the sample and reference, whereas in heat-flux DSC, the thermocouples are attached to individual thermally conducting bases which support the sample and reference holders, resulting in an output signal which is less dependent on the thermal properties of the sample.

In power-compensated DSC, the heat inputs to the sample and the reference are controlled to keep them at the same temperature (i.e. $\Delta T = 0$) throughout the heating programme. The energy difference between the supplies is then plotted against the programme temperature to give a DSC curve. Again, the curves should be marked with either the endo or exo direction to avoid confusion. In DSC, it is common for endothermic responses to appear as positive peaks, i.e. above the baseline, representing an increased transfer of heat to the sample. Unfortunately, this is the reverse of the convention in DTA, where endothermic events appear as downward peaks, representing a negative ΔT, as the temperature of the sample lags behind that of the reference.

12.3.3   Equipment

A thermobalance is a combination of a sensitive microbalance with a furnace and control equipment and a schematic diagram. Null-point balances are preferred for TG since the sample remains in the same region of the furnace throughout the experiment. Any movement of the balance mechanism is detected and a restoring force is applied to maintain the original null position. The force applied is therefore a measurement of mass change. A typical balance will have a sensitivity of about 1 mg and a maximum loading of about 1 g, although this may be increased by altering the balance configuration. The furnace is usually an electrically heated tube, which is raised (or lowered) into position around the sample for thermal measurements. Atmosphere control around the sample is provided by an inert gas purge (this can be replaced by an oxygen purge for combustion or oxidation studies). Solid samples should be powdered (by crushing rather than by grinding, so as to prevent any changes in the surface layers) and be spread thinly and uniformly in the container. Samples which decrepitate (explode) should be covered with a quartz wool plug or fine platinum mesh to prevent material loss from the container. Modern instruments are normally linked to a computer, which

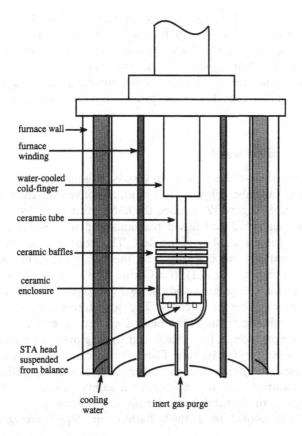

Figure 12.3 Schematic diagram of a simultaneous thermal analyser.

furnace wall

furnace winding

water-cooled cold-finger

ceramic tube

ceramic baffles

ceramic enclosure

STA head suspended from balance

cooling water

inert gas purge

greatly increases the scope for data manipulation in comparison with the X–Y chart output from older instruments.

Simultaneous measurement of TG–DTA or TG–DSC (STA) enables TG to be used as an aid in the interpretation of DTA or DSC results [7,8], and the furnace and sample container configuration for one design is shown in Figure 12.3. Simultaneous IR or mass spectroscopic analysis of the gases evolved during thermal decomposition (EGA) provide a deeper insight into the nature of the chemical changes.

12.4 High pressure reactions

In order to maintain sufficient concentrations of volatile or gaseous reagents during synthetic reactions or catalysis studies, it may be necessary to use a sealed system at greater than atmospheric pressure. Thick-walled glass containers can be used at moderate pressures up to about 7 atm (*ca.* 100 lb inch^{-2}) and have the advantage that the reaction mixture is visible. When higher pressures are required, e.g. in the preparation of metal carbonyls, a metal bomb or autoclave must be used, and designs are available that can withstand pressures of up to about 9000 lb inch^{-2}.

12.4.1 Glass reaction vessels

Of the simpler thick-walled reaction vessels, a screw-top Schlenk flask (Figure 3.6) or tap-tube (Figure 11.1) is more convenient than the traditional Carius tube (Figure 12.4), since the latter must be sealed by a proficient glass blower. Solid and liquid reactants and the solvent are loaded first, together with a stir-bar if needed. The container is then cooled slowly to prevent it from cracking, and the volatile reagents are added from a vacuum line (section 5.3). If a Carius tube is used, it must then be flame-sealed at the constriction and the seal annealed carefully to prevent the glass from cracking as it cools. The sealed reaction vessel is then immersed in a suitable heating bath behind a safety screen. In addition to the usual eye protection, a face shield and thick gloves are recommended for handling pressurized glassware. Carius tubes can also be heated in an oven (the plastic components of the other designs will not withstand the higher temperatures) and, as an additional safety precaution, they may also be enclosed in a metal tube during the heating period. Reaction tubes should be cooled in a slush bath or in liquid nitrogen before

Figure 12.4 Carius tube.

opening, and any gases evolved during the warming period should be vented to a fume cupboard. Condensables from Schlenk or tap-tube reactors can be collected in a cold trap.

Thick-walled glass vessels for use with metal fittings are available commercially and a design for a pressure equalizing dropping funnel for reactions at 50–100 lb inch^{-2} has been described [9].

12.4.2 Metal reaction vessels

Commercial metal reaction vessels are available in a wide variety of configurations, and a bomb with typical fittings is shown in Figure 12.5. In addition to a pressure gauge and a safety rupture disc (with a burst rating to match the range of the gauge), valves for admitting or releasing gas and for removing liquid, a thermocouple and a gas-tight stirrer can be fitted, although not all of these may be necessary. A purpose-built autoclave laboratory or safety barrier enables safe isolation of the bomb assembly and remote control of gas pressures, although microreactors can be used in a fume cupboard with suitable safety screens.

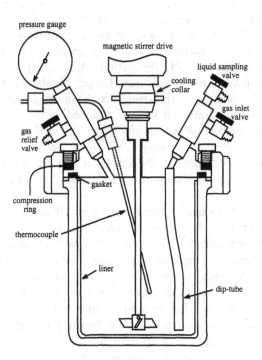

Figure 12.5 Typical stainless-steel pressure reactor.

Choosing a bomb.   Careful thought must be given to the selection of the material of construction and design of the reactor. In addition to several types of stainless steel, a variety of other alloys are available to provide resistance to different chemical environments, and the information and specifications provided by the manufacturers should be consulted. When choosing the reactor type, the main factors to be considered are the volume, the operating temperature and pressure and whether or not stirring is required.

Pressure ratings for reactor designs are given for a particular construction material at a selected temperature, and any increase in temperature above this value will lower the maximum safe operating pressure. Manufacturers provide conversion factors for different temperatures and materials to enable maximum pressure ratings to be calculated.

If stirring is required, magnetically coupled drives provide a system without rotating seals which significantly reduces the chance of leakage under operating conditions. Alternatively, a seal on the rotating stirrer shaft can be provided by packed glands, which require more maintenance than the magnetic drives, although the maintenance frequency will depend on the operating conditions and the nature of the reactor contents.

Reactors must not be filled to more than three-quarters of the free volume, and the most convenient bomb design for investigative chemistry incorporates an O-ring seal and a screw-cap closure with a volume of 25–100 cm^3. In this case, maximum operating temperatures are determined by the O-ring material: 125°C for natural rubber, 220°C for Viton and 275°C for the more expensive Kalrez. Larger bombs have closures with clamping bolts, such as a split-ring which slides on and off without disturbing the connections and fittings attached to the head. Flat Teflon sealing gaskets can be used up to 350°C, while metal or graphite gaskets are suitable for higher temperatures. Glass or Teflon liners supplied by the manufacturers will protect the inside of the reactor from corrosive reaction mixtures, although they also alter the heat transfer characteristics of the bomb.

The simplest way to heat a bomb is to place it directly on a hot plate (most designs have flat bottoms). Alternatively, depending on the size of the bomb, several heater designs are available with automatic temperature control. Band heaters clamp onto the outside of the smaller vessels, rigid or flexible heating mantles are supplied for medium-size vessels, while large capacity vessels are heated by heavy duty heating elements housed in stainless steel shells. Mantles and sheathed-element heaters produce temperatures up to 400°C and heating bands have an upper limit of 500°C, so ceramic heaters must be used for high temperature reactors. A rocking cradle is used with non-stirred pressure vessels to agitate the reaction mixture in a capped glass liner having a breather pinhole. With this arrangement, operating temperatures and pressures in excess of the maximum limits for stirred vessels can be achieved.

Using a bomb. If the starting materials are air-sensitive, smaller bombs may be loaded and assembled in a dry box. Alternatively, the assembled bomb is connected to a manifold *via* the gas release valve and purged with nitrogen while the reaction mixture is prepared in a Schlenk flask. By connecting the liquid sampling valve to a length of wide-bore Teflon tubing which extends into the Schlenk flask through a septum cap, the reaction mixture can then be transferred into the reactor *via* the dip tube as in a normal cannula transfer. Fresh solvent is then added to the flask and transferred into the reactor to wash the valve and dip tube. The valves are then closed and the tightness of the closure checked before placing the bomb in (or on) its heater behind a barrier, where connections to the gas supply, thermocouple and stirrer (if fitted) are made. The gas to be used in the reaction is then added from a cylinder (see section 6.5.1). When the pressure required is greater than that present in the gas cylinder, the gas must be pressurized in a compressor before being admitted to the bomb, and several cycles may be necessary to reach the operating pressure.

After the reaction, the bomb is allowed to cool to room temperature, which may take several hours for large vessels, before gases are vented into a fume cupboard (with care for toxic gases such as carbon monoxide). Products are transferred into a Schlenk flask either in a dry box or through the dip tube by the reverse of the procedure described above for addition of the reaction mixture. Finally, rinse out the bomb with fresh solvent.

Small Teflon-lined digestion bombs are particularly convenient for hydrothermal or 'solvothermal' reactions in near-critical or super-critical solvents (section 12.8). They hold 15–20 cm^3 and the inner Teflon vessel is sealed by tightening a screw cap with a spanner. The maximum operating temperature is 250°C and, since these bombs have no valve fittings, air-sensitive compounds must be loaded into and removed from this type of bomb in a dry box.

12.5 Photochemical reactions

Caution! Ultraviolet radiation is damaging to the eyes and skin and powerful sources can cause blindness in seconds. Wear protective goggles or a face shield, and never look directly at the UV radiation source. The apparatus must be shielded to protect other workers.

UV irradiation is well established as a technique for the generation of highly reactive intermediates, and choices of apparatus for inert-atmosphere synthetic work have been described [10]. Since Pyrex absorbs wavelengths below 320 nm (e.g. only 10% of light with a wavelength of 300 nm is transmitted by a 2 mm thick wall of Pyrex), reactions at lower wavelengths are usually carried out in quartz reaction vessels. A conventional immersion-well reactor for large-scale reactions (10–50 g) is shown in Figure 12.6. A cylindrical medium-pressure mercury arc lamp is contained within a water-cooled quartz tube, which extends into the reaction mixture in a Pyrex reaction flask. The reaction mixture is made up in a separate flask and transferred via cannula to the purged reactor. During the reaction, magnetic stirring is necessary to ensure that the reactants circulate around the immersion well, and a nitrogen purge through the sintered glass inlet at the bottom of the flask removes evolved gases such as CO, driving the reaction towards completion. Since the reaction products may also be sensitive to UV radiation, it is important to monitor the progress of the reaction by removing samples at intervals for IR or NMR spectroscopy. At the end of the reaction, the solution can be filtered through the frit into a Schlenk flask for work-up.

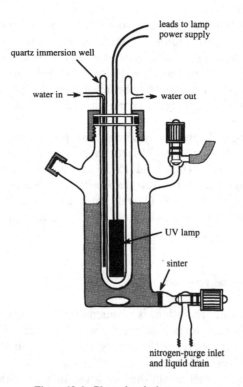

leads to lamp power supply

quartz immersion well

water in →

→ water out

UV lamp

sinter

nitrogen-purge inlet and liquid drain

Figure 12.6 Photochemical reactor.

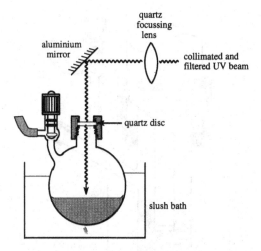

Figure 12.7   Modified Schlenk apparatus for UV photolysis.

For smaller scale reactions (< 5 g) with high quantum efficiencies, it is possible to carry out photochemical reactions in a Schlenk tube with the lower part constructed from quartz tubing and joined to the upper Pyrex tube by a graded seal. In an alternative approach, a collimated and focussed beam of UV light impinges upon the reaction mixture in a standard Schlenk flask through a UV transparent stopper made from a quartz disc (Figure 12.7) [10]. This arrangement enables reactions on a 1–3 g scale to be cooled to temperatures below –30°C in a slush bath (IR radiation is filtered from the beam with distilled water).

12.6   Sonication

Since the mid-1980s, high energy ultrasound has been increasingly used as a means of promoting chemical reactions. Several books [11] and reviews [12] have been published and provide detailed coverage of the area, ranging from introductory to advanced research level, and including applications in synthesis. The basic principles of sonochemistry and the apparatus required for inert-atmosphere reactions are described in this section, in addition to a brief summary of the types of reaction in which sonication has proved to be beneficial.

Many of the developments in non-aqueous systems have resulted from studies by Suslick on homogeneous organometallic reactions, and his design for glass sonochemical reaction vessels (Figure 12.8) has been widely adopted by other research groups [13]. An ultrasonic horn, of the

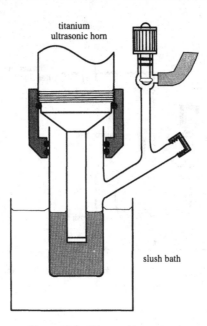

titanium
ultrasonic horn

slush bath

Figure 12.8 Ultrasonic reactor.

type used by biochemists for cell disruption, amplifies the output of a transducer and is held in the reaction vessel by a threaded stainless steel or Teflon collar fitted with O-ring seals. In order not to dampen the vibration of the probe tip, the horn is threaded at a node in the acoustic standing wave set up in the probe. The temperature inside the reaction vessel during sonication should be monitored with a thermocouple, since any significant temperature rise will increase the vapour pressure of the solvent, which has been found to reduce the intensity of acoustic cavitation. Hence, the reaction vessel is immersed in a cooling bath.

It is generally accepted that most chemical effects of ultrasound are due to the formation and collapse of microbubbles in the liquid, a process known as cavitation. This generates local transients of very high temperature and pressure which can affect the course of homogeneous and heterogeneous reactions. Sonication provides an efficient method of accelerating many reactions involving metal surfaces, and while some of these can be carried out using an ultrasonic cleaning bath, most will benefit from the use of an immersion reactor. Thus, Grignard reactions are easier to initiate and reactions with zinc, lithium or copper occur under mild conditions. Reactive metal powders (section 6.4.4) can also be generated by sonication [14]. In general, the reduction in particle size upon sonication enhances reactivity in heterogeneous reactions involving solids. Similar effects are observed in reactions between immiscible liquids, where sonication generates very fine emulsions (in some cases, further benefit may be gained by the addition of a phase-transfer catalyst).

When operating a high power ultrasonic reactor, it is advisable to wear ear protection and to site the reactor away from other workers, preferably in a separate room.

12.7 Microwave heating

Heating effects arising from the interaction of microwave radiation with the dielectric properties of materials have found increasing use in chemical synthesis over the last 10 years or so. Domestic microwave ovens, operating at a frequency of 2.45 GHz and producing around 600 W of electromagnetic energy, can be used with or without modification for synthetic applications. Solvents with low molecular weights and high dipole moments are heated by microwave radiation, and water, methanol, ethanol, acetonitrile, N,N-dimethylformamide and dichloromethane are typical examples, while reactions in non-polar solvents such as arenes or hydrocarbons require the addition of a polar solvent to provide coupling to the microwaves. The high voltages used in microwave ovens may cause sparks, so flammable solvents must be properly contained to prevent an explosion in the microwave cavity.

A conventional microwave oven must be modified if reactions are to be carried out at atmospheric pressure under reflux, and the arrangement shown in Figure 12.9 is that described by Mingos and Baghurst [15]. The remote condenser is connected to the flask by a tube which protrudes through a port in the oven wall. A Teflon inlet tube for nitrogen is also fed through the wall and connected to the flask. To prevent microwave leakage from the oven, the tubes are enclosed in flanged copper tubes which are bolted to the side of the oven to ensure good electrical contact.

Alternatively, reactants can be heated in a Parr digestion bomb that will withstand temperatures up to 250°C and pressures up to 80 atm (see section 12.4.2) [16]. Constructed from microwave-transparent polymers with an inner Teflon reaction vessel, the reactor can be placed directly in a conventional microwave oven. The temperature inside the bomb rises to about 200°C in less than a minute, so a pressure-relief disc is incorporated to release excess pressures. This heating method has proved useful in the preparation of a range of organometallic and coordination compounds, especially when the metal is substitutionally inert [15], and in hydrothermal synthesis of inorganic materials [17]. Microwave heating of reactions between solids has been shown to be a convenient method of synthesizing oxide, sulphide and nitride materials, provided at least one of the reactants couples with the microwave radiation. The reactants are

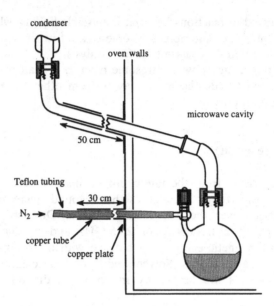

Figure 12.9 Reactions in a modified microwave oven.

sealed in a quartz tube, which is then placed in the microwave oven. Finely divided metal powders may also be used in these reactions, since the high surface area and small particle size prevents the charge build-up which would otherwise result in sparking [18].

12.8 Reactions in special solvents

In certain circumstances, the choice of an exotic solvent or the use of a near- or super-critical solvent may provide the optimum conditions for a reaction. This section outlines the advantages to be gained by these practices and describes examples of their use, emphasizing the importance of innovation in practical chemistry.

12.8.1 Condensable gases

Liquid ammonia. Although liquid ammonia (b.p. −33°C) is not considered a very exotic solvent, the techniques involved in its use are generally applicable to other condensable gases. **Caution!** Ammonia is an irritating, toxic and flammable gas. Carry out all operations in an efficient fume cupboard, wear rubber or plastic gloves and avoid contact with the liquid.

It is best to transfer the ammonia from a cylinder as a liquid (section

6.5, Figure 6.9) into a flask cooled in a slush bath, where it can be used directly if the reaction does not require dry conditions. Traces of water can be removed by adding a couple of lumps of sodium to the flask to generate a blue solution and then distilling the ammonia by vacuum transfer (section 5.3) through a connector containing glass wool and a sinter (to retain sodium) into a clean flask containing any solid reagents and a stirrer (Figure 12.10a). A cold-finger condenser and oil bubbler are then fitted to the flask and any remaining liquid reagents are added (Figure 12.10b). For reactions at low temperature, or if a long reaction time is anticipated, immerse the flask in a cooling bath, otherwise allow the ammonia to reflux at room temperature until the reaction is complete. Products are isolated by allowing the ammonia to evaporate at room temperature (in the fume cupboard).

Liquid ammonia is most commonly used in 'dissolved metal' reactions (e.g. Birch reductions) and for the preparation of sodium and lithium amide. For example, $NaPPh_2$ is readily prepared from sodium metal and PPh_3 in liquid ammonia.

Liquid sulphur dioxide. Although less frequently used than liquid ammonia, this is a useful reactant and low-temperature solvent (b.p. $-10°C$) and can be handled on a vacuum line in the normal fashion (section 5.3). Room temperature electrochemical measurements have been carried out in liquid SO_2 using a cell and reference electrode that can withstand pressures up to 3 atm [19].

12.8.2 Super-critical fluids

The properties of the super-critical phase make it attractive for preparative chemistry, and hydrothermal techniques have been used for over 150 years for the synthesis of inorganic solids in aqueous solutions. Reactants that are poorly soluble under ambient conditions will often dissolve under hydrothermal conditions, and low viscosities provide good conditions for crystal growth. The review by Rabenau provides a useful introduction to the methodology [20].

'Hydrothermal synthesis' generally refers to reactions in aqueous media at temperatures above $100°C$ and pressures above 1 atm, and no distinction is made between reactions below and those above the critical point. In recent years, a renewed interest in non-aqueous media for 'solvothermal' reactions has led to the preparation of new inorganic and organometallic compounds, e.g. super-critical amine solvents have enabled the preparation of new ternary and quaternary compounds [21]. Near-critical conditions provide a combination of liquid-like solvent properties with gas-like transport properties, and the reduced viscosities (which can be up to 100 times smaller than under normal conditions) result in

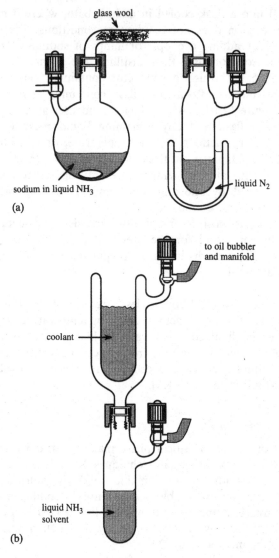

glass wool

liquid N$_2$

sodium in liquid NH$_3$

(a)

to oil bubbler
and manifold

coolant

liquid NH$_3$
solvent

(b)

Figure 12.10 (a) Drying and transfer of liquid ammonia. (b) Reactions in liquid ammonia under reflux.

higher rates of diffusion-controlled reactions. Critical temperatures (T_c), pressures (P_c) and molar volumes (V_c) for selected solvents are listed in Appendix C. Note that the presence of a solute in a solvent will usually shift the critical point to higher temperatures.

Sealed, thick-walled borosilicate or quartz glass ampoules are convenient containers for neutral or acid mixtures at temperatures up to

250–300°C or about 500°C, respectively, and will withstand pressures of several hundred atmospheres. These have the advantage that the reaction mixture is visible, but precautions must be taken in case the ampoule bursts. At higher pressures, the internal pressure is compensated by placing the ampoule inside an autoclave which is then pressurized with inert gas.

In many cases, it is simpler to use a metal digestion bomb with a Teflon liner and a safety rupture disc (section 12.4.2). The maximum working pressure of a Parr general purpose digestion bomb with a volume of 23 cm^3 is 1800 lb inch^{-2} (122 atm), while a high-pressure version can be obtained which will withstand 5000 lb inch^{-2} (340 atm).

The pressure in a reaction vessel is determined by the percentage of the volume initially occupied by the solvent at ambient conditions, and the percentage fill necessary for critical conditions is determined from the critical molar volume V_c. For example, V_c for water is 56 cm^3 mol^{-1}, i.e. 18 g of pure water will occupy 56 cm^3 at T_c and P_c, requiring the reactor to be at least 32% filled for normal hydrothermal conditions. Given that the presence of the reactants will affect the critical points, near-critical conditions are likely to be achieved by using a temperature of at least T_c and a minimum volume of solvent determined from V_c. The amount of material that can be prepared using these reactors is limited by their volume, and scale-up to larger high-pressure bombs has significant cost and safety implications.

A significant development by Poliakoff and co-workers has been the design of miniature flow reactors to overcome restrictions imposed by large-scale, high-pressure bombs and to allow *in situ* monitoring of UV photolysis reactions [22]. Super-critical fluid containing low concentrations of metal carbonyls (to minimize the effect on critical points) is passed through a small-volume UV photolysis cell, an IR cell and a pressure regulator before the organometallic product is collected. Using this process, [Cr(CO)$_5$(C$_2$H$_4$)] was prepared in super-critical ethene and [Mn(η^5-C$_5$H$_5$)(CO)$_2$(η^2-H$_2$)] was obtained from a super-critical hydrogen/carbon dioxide mixture.

12.8.3 Ambient temperature ionic liquids

Anhydrous haloaluminate salts formed by addition of Q$^+$X$^-$ to AlX$_3$ (usually where the organic cation Q$^+$ is 1-methyl, 3-ethylimidazolium) are molten at room temperature and have attracted attention because they will dissolve organic and inorganic compounds, and can stabilize novel coordination species [23]. The properties of these melts can be varied from basic to acidic (i.e. from halide donors to halide acceptors) by changing the molar ratio of AlX$_3$ to Q$^+$X$^-$ from below 1 to a value between 1 and 2, which affects the relative amounts of [AlX$_4$]$^-$ and

$[Al_2X_7]^-$. The melts are moisture-sensitive and the quaternary organic halide and the aluminium halide should be carefully purified before use. Traces of oxide materials can be removed from the molten salt by treatment with phosgene ($COCl_2$) (note that phosgene is very toxic and stringent safety procedures must be adopted) [24]. Electrochemical studies of metal halide clusters in chloroaluminate melts provide a nice illustration of the potential of these alternative solvents [25].

12.9   Matrix isolation

To a preparative chemist, cryogenic matrix isolation may seem an exotic technique more suited to theoretical spectroscopists. However, its application to the study of organometallic photochemistry has revealed mechanistic details of important reactions, including C–H bond activation [26]. In addition, matrix isolation studies may also reveal fruitful avenues for synthetic investigations. This section has been included to provide a brief introduction to the technique, while technical details of matrix isolation experiments are described in the references.

By co-condensing a volatile sample with an excess of a gas onto the windows of a spectroscopic cell at low temperatures (4–20 K), the sample is dispersed within a solid matrix and can be studied by conventional spectroscopic methods. Possible matrix gases are Ne, Ar, Kr, Xe, CH_4, N_2 and CO. At one extreme, Ne is totally inert, while at the other, CO may react with many guest molecules. Samples may be gaseous reaction products for analysis, or the starting materials for *in situ* photochemical reactions.

A relatively inexpensive demonstration of the photochemical generation of $Cr(CO)_5$ in a mineral-oil matrix cooled by liquid nitrogen has been described [27], whereas research equipment is much more expensive, requiring a helium cryostat or refrigerator and a high vacuum system, in addition to a photolysis source and high-resolution spectrometer. A schematic diagram of a basic cryostat is shown in Figure 12.11. The coolant reservoir is attached to a metal cryotip which holds the sample window inside a vacuum jacket. In this design, the window is rotated from the inert-gas and sample inlet to a position for photolysis with a laser or mercury arc lamp (with appropriate filters) and analysis with IR or UV/visible radiation. More complex optical systems allow time-resolved spectroscopic studies. CsI windows are suitable for IR studies, while BaF_2 is used for combined IR and UV/Vis detection.

Matrix isolation has proved particularly useful for studying the interactions between metal atoms and potential ligands at low temperatures, and is often used to complement synthetic studies. The metal atoms are

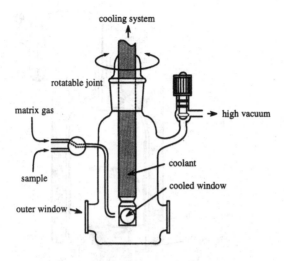

Figure 12.11 Simple matrix isolation apparatus.

evaporated from a small furnace attached to the cell and condensed onto the cold window together with the reactant and an excess of the matrix gas. Methods for carrying out metal atom reactions on a preparative scale are described in the following section.

12.10   Metal vapour synthesis

The condensation of metal atoms with potential ligands continues to be a rich source of new compounds that are unobtainable or difficult to prepare by normal procedures [28]. The synthesis of organometallic compounds from metal vapours has been reviewed [29], and recent developments have been described by Timms [30]. The basic requirements of a metal vapour synthesis (MVS) reactor are a source of high vacuum, a furnace from which the metal can be evaporated and a cold surface onto which the metal vapour and the reactants are condensed. Different designs of apparatus have evolved from the work of several research groups and equipment is available commercially [31].

12.10.1   Heating options

A crucible constructed from a spiral of molybdenum or tungsten wire and covered with alumina paste is suitable for evaporating Cr, Mn, Fe, Co, Ni, Cu, Pd, Ag, Au, and other metals that do not attack alumina. Some

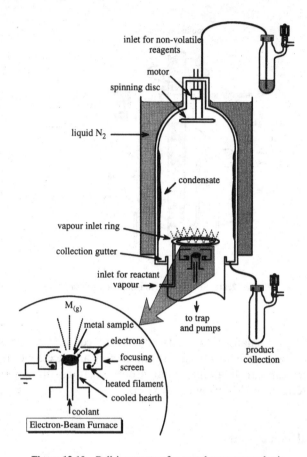

Figure 12.12 Bell-jar reactor for metal vapour synthesis.

lanthanides, V and Cr may be evaporated from a resistance-heated Ta or W boat, while Ti, Mo or W may be sublimed on a small scale from resistively heated wires. However, resistance heating is generally not suitable for the refractory metals and the thermal radiation produced may cause problems. Consequently, electron-beam heating has been adopted as the preferred method for evaporation of metals at rates of 1–2 g h^{-1}. In an electron-beam furnace, electrons emitted from a heated filament are focused onto a lump of metal supported on a cooled hearth. The emission of secondary electrons from the metal is minimized by operating the electron-beam furnace in a reversed-polarity mode, i.e. with the metal and hearth at positive potential [32].

12.10.2 Static reactors

The most straightforward MVS reactor design involves a glass or stainless-steel bell-jar cooled by liquid nitrogen, as shown in Figure 12.12 with an electron-beam furnace. The metal is first degassed in a preliminary step by placing it on the hearth of the electron gun, evacuating the uncooled bell-jar to about 10^{-5} mmHg and heating the sample slowly until it melts. Pumping is continued throughout the outgassing process until the original pressure is regained.

After the pre-melted bead of metal has been placed on the furnace, the apparatus is assembled, evacuated to better than 10^{-5} mmHg and cooled with liquid nitrogen. The furnace power is then increased slowly to give the required evaporation rate. Volatile reactants are introduced through a perforated inlet ring from a reservoir *via* a needle valve and heated manifold and condense along with the metal atoms on the reactor wall. After the required time (between 1 and 4 h, depending on the evaporation rate), the furnace is turned off and the reactor is allowed to warm to room temperature under argon (nitrogen may interact with the low-valent metal complexes formed under these conditions). When the solid matrix has melted, the liquid which collects at the bottom of the reactor is filtered immediately into a Schlenk flask to remove any metal particles, which may cause decomposition of the products.

Involatile reagents can be introduced as a neat liquid or as a solution in an inert solvent, and the dispersion of fine droplets with a spinning disc to

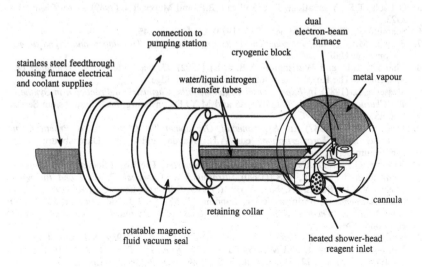

Figure 12.13 Rotary reactor for metal vapour synthesis.

give a falling liquid film on the wall of the reactor has been described [32]. The higher vapour pressures in this configuration prevent the use of the electron-beam furnace unless the area around the hearth is cryopumped by cooling with liquid nitrogen, thus maintaining a local pressure of less than about 5×10^{-4} mmHg.

12.10.3 Rotating reactors

An alternative reactor design for liquid-phase reactions with metal vapours has been described (Figure 12.13) [31a,33]. The reaction flask is rotated during operation, the high vacuum being maintained by a magnetic fluid seal. Cryopumping enables an electron-beam furnace to be used and the rate of metal evaporation is monitored with a quartz crystal microbalance. Liquid reactants are introduced through a heated shower head and form a band around the wall of the rotating flask, which is cooled in a slush bath during the reaction. Products are removed through stainless-steel cannulae.

References

1. Christensen, P.A. and Hamnett, A. (1994) *Techniques and Mechanisms in Electrochemistry*, Blackie, Glasgow.
2. Greef, R., Peat, R., Peter, L.M., *et al.* (1990) *Instrumental Methods in Electrochemistry*, Ellis Horwood, Chichester.
3. Sawyer, D.T. and Roberts, Jr., J.L.(1974) *Experimental Electrochemistry for Chemists*, John Wiley and Sons, New York.
4. Mabbott, G.A. (1983) *J. Chem. Educ.*, **60**, 697.
5. O'Toole, T.R., Younathan, J.N., Sullivan, B.P. and Meyer, T.J. (1989) *Inorg. Chem.*, **28**, 3923.
6. Schmulbach, C.D. and Oomen, T.V. (1973) *Anal. Chem.*, **45**, 820.
7. Brown, M.E. (1988) *Introduction to Thermal Analysis: Techniques and Applications*, Chapman & Hall, London.
8. Charsley, E.L. and Warrington, S.B. (eds) (1992) *Thermal Analysis – Techniques and Applications* The Royal Society of Chemistry, Cambridge.
9. Messerle, L. (1987) In *Experimental Organometallic Chemistry: a Practicum in Synthesis and Characterization*, eds A.L. Wayda and M.Y. Darensbourg, ACS Symposium Series, Vol. 357, p. 199.
10. Trogler, W.C. (1987) In *Experimental Organometallic Chemistry: a Practicum in Synthesis and Characterization*, eds A.L. Wayda and M.Y. Darensbourg, ACS Symposium Series, Vol. 357, p. 70.
11. (a) Price, G.J. (1992) *Current Trends in Sonochemistry*, The Royal Society of Chemistry, Cambridge; (b) Suslick, K.S. (1990) *Ultrasound: its Chemical, Physical and Biological Effects*, V.C.H. Publishers, New York; (c) Ley, S.V. and Low, C.M.R. (1989) *Ultrasound in Chemistry*, Springer Verlag, London; (d) Mason, T.J. and Lorimer, J.P. (1989) *Sonochemistry: Theory, Applications and Uses of Ultrasound in Chemistry*, Ellis Horwood, Chichester.
12. (a) Suslick, K.S. (1990) *Science*, **247**, 1439; (b) Suslick, K.S. (1989) *Scientific American*, **260**, 80; (c) Lindley, J. and Mason, T.J. (1987) *Chem. Soc. Rev.*, **16**, 275; (d) Bremner, D. (1986) *Chem. Br.*, **22**, 633; Suslick, K.S. (1986) *Adv. Organomet. Chem.*, **25**, 73.
13. Suslick, K.S. and Flint, E.B. (1987) In *Experimental Organometallic Chemistry: a*

Practicum in Synthesis and Characterization, eds A.L. Wayda and M.Y. Darensbourg, ACS Symposium Series, Vol. 357, p. 195.
14. Fürstner, A. (1993) *Angew. Chem. Int. Ed. Eng.*, **32**, 164.
15. Mingos, D.M.P. and Baghurst, D.R. (1991) *Chem. Soc. Rev.*, **20**, 1.
16. Parr Instrument Company, 211, 53rd St, Moline, IL 61265, USA. UK distributors: Scientific and Medical Products Ltd, Shirley Institute, 856 Wilmslow Road, Didsbury, Manchester, M20 8RX.
17. Baghurst, D.R., Barrett, J. and Mingos, D.M.P. (1995) *J. Chem. Soc., Chem. Commun.*, 323.
18. Mingos, D.M.P. (1993) *Advanced Materials*, **5**, 857.
19. Banister, A.J., Hauptman, Z.V., Kendrick, A.G. and Small, R.W.H. (1987) *J. Chem. Soc., Dalton Trans.*, 915; Aherne, C.M., Banister, A.J., Gorell, I.B., Hansford, M.I., Hauptman, Z.V., Luke, A.W. and Rawson, J.M. (1993) *ibid.*, 967.
20. Rabenau, A. (1985) *Angew. Chem. Int. Ed. Eng.*, **24**, 1026.
21. Jerome, J.E., Wood, P.T., Pennington, W.T. and Kolis, J.W. (1994) *Inorg. Chem.*, **33**, 1733.
22. Banister, J.A., Lee, P.D. and Poliakoff, M. (1995) *Organometallics*, **14**, 3876.
23. Hussey, C.L. (1988) *Pure Appl. Chem.*, **60**, 1763.
24. Abdul-Sada, A.K., Avent, A.G., Parkington, M.J., *et al.* (1993) *J. Chem. Soc., Dalton Trans.*, 3283 (this article contains details of the precautions that are necessary for handling phosgene gas).
25. Hussey, C.L., Quigley, R. and Seddon, K.R. (1995) *Inorg. Chem.*, **34**, 370.
26. Fausto, R. (1996) *Low Temperature Spectroscopy*, Kluwer Academic Publishers, Dordrecht; Perutz, R.N. (1994) *Chem. Soc. Rev.*, **22**, 361; Grebenik, P., Grinter R. and Perutz, R.N. (1988) *ibid.*, **17**, 453.
27. Flair, M. and Fletcher, T.R. (1995) *J. Chem. Educ.*, **72**, 753.
28. Cloke, F.G.N. (1993) *Chem. Soc. Rev.*, **22**, 17.
29. Timms, P.L. and Turney, T.W. (1977) *Adv. Organomet. Chem.*, **15**, 53; Green, M.L.H. (1980) *J. Organomet. Chem.*, **200**, 119.
30. Timms, P.L.(1996) *Chem. Soc. Rev.*, **25**, 93.
31. (a) Andrews, M.P. (1987) In *Experimental Organometallic Chemistry: a Practicum in Synthesis and Characterization*, eds A.L. Wayda and M.Y. Darensbourg, ACS Symposium Series, Vol. 357, p. 158; (b) Groschens, T.J. and Klabunde, K.J. (1987) *ibid.*, p. 190 and p. 193.
32. Cloke, F.G.N. and Green, M.L.H. (1981) *J. Chem. Soc., Dalton Trans.*, 1938.
33. Ozin, G.A., Andrews, M.P., Francis, C.G., *et al.* (1990) *Inorg. Chem.*, **29**, 1068.

13 Preparation of Starting Materials

13.1 Introduction

The following sections contain references to, and very brief descriptions of, preparative methods for selected compounds. After some deliberation, I decided not to reproduce detailed descriptions of syntheses (this is, after all, why *Inorganic Syntheses* is published) but rather to provide rapid access to published syntheses of compounds which may be required as starting materials. The selection is inevitably subjective, and reflects my own research interests, but I have aimed to cover a wide range of metals and oxidation states. Wherever possible, tested procedures from *Inorganic Syntheses* are given, and I have abbreviated these references to (**volume**, page), but other references are also included where appropriate. It is worth noting that there are cumulative indices in volumes **10, 15, 20, 25,** and **29** of *Inorg. Synth.*, and that volumes **28** and **29** contain compilations of starting compounds. Armed with the information provided in the preceding chapters of this book, you should be familiar with the techniques described in the articles and (hopefully) be able to repeat the syntheses. When searching for a suitable preparative method for a given compound, it is also worth checking the *Dictionary of Inorganic Compounds*, the *Dictionary of Organometallic Compounds* and the *Dictionary of Organic Compounds* published by Chapman and Hall before searching *Chem. Abs.* or *Gmelin*.

In each of the sections below, the metal compounds are listed in order of increasing group number (and then period within each group) for the metal associated with the ligand of interest (i.e. discounting counter-ions).

Safety note: before attempting any of these syntheses, consult the references for details of hazards and safety precautions.

13.2 Halides and their simple derivatives

Metal halides are frequently the starting point for syntheses, and in cases where the compound is not available commercially, or where the commercial product is insufficiently pure or too expensive, it may be necessary to prepare or purify your own material. The range of procedures given below include simple dehydration of commercially available hydrated compounds.

13.2.1 M(I) halides

LiCl (**28**, 321)

Dehydration of hydrated samples by treatment with $SOCl_2$.

$RhCl(PPh_3)_3$ (**28**, 77)

Treatment of $RhCl_3 \cdot xH_2O$ in ethanol with a solution of PPh_3 in hot ethanol followed by 2 h reflux. Maroon crystalline product, or orange 'isomeric' form if smaller volumes or shorter reflux times are used.

$RhCl(CO)(PPh_3)_2$ (**28**, 79)

Addition of an ethanolic solution of $RhCl_3 \cdot xH_2O$ to a boiling ethanolic solution of PPh_3 and then addition of formaldehyde solution to give the yellow microcrystalline product.

$[RhCl(CO)_2]_2$ (**28**, 84)

Pulverized $RhCl_3 \cdot xH_2O$ is heated in a CO stream at 100°C. Orange crystals sublime in reaction vessel.

$[RhCl(C_2H_4)_2]_2$ (**28**, 86)

Aqueous solution of $RhCl_3 \cdot xH_2O$ transferred to MeOH. Degassed under reduced pressure, then stirred under 1 atm pressure of C_2H_4 to give orange precipitate after about 1 h. Product collected after about 7 h.

$[RhCl(C_8H_{12})]_2$ (**28**, 88)

Treatment of $RhCl_3 \cdot xH_2O$ in ethanol with cyclooctadiene, 18 h reflux. Na_2CO_3 can be added if the starting chloride is acidic. Yellow–orange, air-stable solid.

$[RhCl(C_8H_{14})_2]_2$ (**28**, 90)

$RhCl_3 \cdot xH_2O$ in Pr^iOH/H_2O stirred with cyclooctene for 15 min, then allowed to stand at room temp. for 5 days. Orange–yellow solid.

$[IrCl(C_8H_{14})_2]_2$ (**28**, 91)

Suspension of a $[IrCl_6]^{3-}$ salt (ammonium, sodium or potassium) in Pr^iOH/H_2O treated with cyclooctene and heated under reflux for 3–4 h. Oily orange product crystallizes under ethanol at 0°C. Yellow crystals.

$IrCl(CO)(PPh_3)_2$ (**28**, 92)

$IrCl_3 \cdot xH_2O$, PPh_3 and aniline in DMF heated to reflux for 12 h. Addition of methanol and cooling gives yellow crystals.

$AuCl(PPh_3)$ (**26**, 325; **27**, 218)

$HAuCl_4 \cdot xH_2O$ treated with PPh_3 in ethanol.

13.2.2 M(II) halides

$BaCl_2$, $CoCl_2$, $CuCl_2$, $ZnCl_2$ (**29**, 110)

Dehydration of hydrated salts with Me_3SiCl in THF (neat Me_3SiCl required for $CoCl_2$).

CoCl$_2$, NiCl$_2$, CuCl$_2$, ZnCl$_2$, CdCl$_2$ (**28**, 321)

Dehydration of hydrated salts with SOCl$_2$.

LaI$_2$ (**30**, 17)

Solid-state reaction between LaI$_3$ and La in sealed metal (Ta, Mo or W) container. Good reducing agent. Store in absence of O$_2$ and H$_2$O.

TiCl$_2$(TMEDA) [1]

TiCl$_3$(THF)$_3$ in THF stirred with Li and an excess of TMEDA. Pale violet crystals.

VCl$_2$ (**4**, 126)

Reduction of VCl$_3$ with H$_2$ at 400–675°C. Green crystals.

[V$_2$Cl$_3$(THF)$_6$]$_2$[Zn$_2$Cl$_6$] (**28**, 263)

Reduction of VCl$_3$ with Zn dust in THF. Green crystalline powder, useful for synthesis of organometallic V compounds.

CrX$_2$(H$_2$O)$_n$ (X = Cl, n = 4; X = Br, I, n = 6) (**10**, 27)

From Cr metal and the aqueous acid under N$_2$. Bright blue crystals on cooling.

CrCl$_2$(MeCN)$_2$ (**10**, 31)

CrCl$_2$(H$_2$O)$_4$ heated at 140°C *in vacuo*. The resulting anhydrous CrCl$_2$ is dissolved in EtOH and MeCN added to give pale blue crystals.

Mo$_6$Cl$_{12}$ (**12**, 172)

Solid-state reaction between MoCl$_5$ and Mo in a horizontal reactor. Yellow powder.

Mo$_6$Br$_{12}$ (**12**, 176)

Solid-state reaction between Mo$_6$Cl$_{12}$ and LiBr·xH$_2$O.

W$_6$Cl$_{12}$ (**30**, 1)

Solid-state reaction between WCl$_6$ and W in a sealed silica tube.

WCl$_2$(PR$_3$)$_4$ (**28**, 329)

WCl$_4$(PR$_3$)$_2$ reduced with Na/Hg amalgam in THF or toluene. Orange crystals.

RuCl$_2$(PPh$_3$)$_4$ (**12**, 238)

RuCl$_3$·xH$_2$O in MeOH, filtered, heated to reflux under N$_2$, cooled and PPh$_3$ added. Shaken for 24 h to give brown crystals.

RuCl$_2$(PPh$_3$)$_3$ (**12**, 238)

Similar to above, but reaction mixture refluxed for 3 h after addition of phosphine. Black crystals.

NiCl$_2$·2H$_2$O (**13**, 156)

NiX$_2$·6H$_2$O heated to 80°C in an oven purged with air.

NiBr$_2$·2H$_2$O (**13**, 156)

From NiO and aqueous HBr. Product isolated from hot solution as yellow–brown crystals.

NiCl$_2$(DME)$_2$ (**13**, 160)

Slurry of $NiCl_2\cdot 2H_2O$ in DME and $HC(OEt)_3$ heated under reflux. Orange solid.

$NiBr_2(DME)_2$ (X = Cl, Br) (13, 162)

Slurry of $NiBr_2\cdot 2H_2O$ in EtOH and $HC(OEt)_3$ heated under reflux. Solution concentrated and DME added. Salmon-pink solid.

$NiI_2(DME)_2$ (13, 163)

$NiI_2\cdot 5H_2O$ added to $HC(OEt)_3$, stirred for 3 h and volatiles removed *in vacuo*. Residue recrystallized from hot DME. Orange–red crystals.

$PdCl_2(PhCN)_2$ (28, 61)

$PdCl_2$ in PhCN heated to 100°C. Precipitated by adding solution to petrol. Yellow–orange solid. Readily converted to $PdCl_2L_2$ complexes.

$PdCl_2(C_8H_{12})$ (28, 348)

$PdCl_2$ dissolved in conc. HCl. Solution diluted with EtOH, filtered, and then treated with 1,5-cyclooctadiene. Precipitate recrystallized from CH_2Cl_2. Yellow crystals.

$PtCl_2$ (20, 48)

Thin layer of $H_2PtCl_6\cdot xH_2O$ in a combustion boat. Furnace raised to 350°C in 50°C steps over 3 h then held at 350°C for 30 min. Brown solid.

$PtCl_2(PhCN)_2$ (28, 62)

$PtCl_2$ added to hot PhCN. Filtered and precipitated with petrol. Pale yellow solid.

13.2.3 M(III) halides

$ScCl_3(THF)_3$ (21, 139)

Dehydration of hydrated salt with $SOCl_2$ in THF.

LaI_3 (30, 11)

Reaction between La metal and HgI_2 in a pyrex tube at 300–330°C for 12–48 h, or from La metal and I_2 in a sealed metal tube.

$LnCl_3(THF)_n$ (Ln = Yb, Er, Sm, Nd) (28, 286)

Reaction between an excess of Ln metal powder and $HgCl_2$ in THF.

$TiBr_3$ (26, 382)

Reaction between $TiBr_4$ and Ti in a sealed silica ampoule. Heated to 500°C at 50°C h^{-1} then at 500°C for 15 days.

$TiCl_3(THF)_3$ (21, 137)

Solution of $TiCl_3$ in THF refluxed for 22 h. Pale blue crystals on cooling.

VCl_3 (4, 128)

Thermal decomposition of VCl_4 at 160–170°C over 50 h. Violet solid.

$VCl_3(THF)_3$ (**21**, 137)

Solution of VCl_3 in THF refluxed for 22 h.

$NbX_3(DME)$ (X = Cl, Br) (**29**, 120, 122)

Reduction of NbX_5 with Bu^n_3SnH in DME at low temperature. $NbCl_3(DME)$ is a brick-red solid and $NbBr_3(DME)$ is a purple solid.

$CrCl_3(THF)_3$ (**29**, 110)

Dehydration of hydrated salt with Me_3SiCl in THF. Purple solid.

$MoBr_3$ (**10**, 50)

Reaction between Mo and Br_2 in a sealed tube.

$MoCl_3$ (**21**, 51)

Reaction between $Mo(CO)_6$ and $MoCl_5$ in refluxing chlorobenzene.

$MoCl_3(THF)_3$ (**28**, 36)

Reduction of $MoCl_4(THF)_2$ with Sn powder in THF. Pale orange crystals.

$MoCl_3(MeCN)_3$ (**28**, 37)

Reduction of $MoCl_4(MeCN)_2$ with Sn powder in MeCN. Microcrystalline yellow powder, soluble in CH_2Cl_2 but only sparingly soluble in MeCN. MeCN ligands more labile than THF.

$[NH_4]_3[MoCl_6]$ (**29**, 127)

$(NH_4)_6Mo_7O_{24} \cdot 4H_2O$ dissolved in conc. HCl and reduced with Sn. Red crystals.

$NaW_2Cl_7(THF)_5$

Reduction of WCl_4 with Na/Hg amalgam in THF [2]. Green crystals. Originally formulated as $W_2Cl_6(THF)_4$ [3].

Re_3Cl_9 (**20**, 44)

Thermal decomposition of refluxing $ReCl_5$. Dark red crystals.

$[Bu^n_4N]_2[Re_2Cl_8]$ (**28**, 332)

$[Bu^n_4N]ReO_4$ treated with benzoyl chloride under reflux at slightly elevated pressure (so that b.p. = 209°C) for 1.5 h. Solution treated with an ethanolic solution of $[Bu^n_4N]Br$ which has been saturated with HCl and refluxed for further 1 h. Blue crystals.

$FeCl_3$ (**29**, 110)

Dehydration of hydrated salt with neat Me_3SiCl.

$[Et_4N][RuCl_4(MeCN)_2]$ (**26**, 356)

$RuCl_3 \cdot xH_2O$ in conc. HCl refluxed for 10 h. $[Et_4N]Cl \cdot xH_2O$ and Hg added to cooled solution, which is then shaken for 3 h. Filtered solution suspended in MeCN and refluxed for 10 h. Air-stable yellow solid.

$HAuCl_4 \cdot xH_2O$ (**4**, 14)

Au dissolved in a mixture of conc. HNO_3 and conc. HCl, volume reduced and more conc. HCl added.

13.2.4 M(IV) halides

$MCl_4(THF)_2$ (M = Ti, Zr, Hf) (**21**, 137)
THF added to MCl_4 in CH_2Cl_2.

TiI_4 (**10**, 1)
Reaction between Ti metal and I_2 vapour at 400–425°C (horizontal reactor).

VCl_4 (**20**, 41)
Reaction between V and Cl_2 in a vertical reactor.

$NbCl_4(MeCN)_2$ (**21**, 138)
Reduction of $NbCl_5$ with Al in MeCN. Pale yellow powder.

$NbCl_4(THF)_2$ (**29**, 120)
Reduction of $NbCl_5$ with Bu^n_3SnH in toluene/THF. Yellow powder.

$NbBr_4(THF)_2$ (**29**, 121)
Reduction of $NbBr_5$ with Bu^n_3SnH in toluene/THF. Red–orange powder.

$MoBr_4$ (**10**, 49)
Reaction between $MoBr_3$ and Br_2 in a sealed tube.

$MoCl_4(MeCN)_2$ (**28**, 34)
Reduction of $MoCl_5$ with an excess of MeCN. Orange–brown powder.

$MoCl_4(THF)_2$(**28**, 35)
Treatment of $MoCl_4(MeCN)_2$ with THF. Orange–yellow powder.

WCl_4 (**29**, 138)
Solid-state reaction between WCl_6 and red phosphorus in a sealed tube. Alternatively, from the reaction between WCl_6 and $W(CO)_6$ in chlorobenzene (**26**, 221).

$WCl_4(PR_3)_2$ (**28**, 326)
From WCl_4 and PR_3 or by reduction of WCl_6 with an excess of PR_3.

UCl_4 (**21**, 187)
UO_2 prepared *in situ* from uranyl oxalate and treated with CCl_4 in a H_2 flow. Dark green solid.

13.2.5 M(V) halides

MCl_5 (M = Nb, Ta, Re) (**20**, 41)
Reaction between the metal and Cl_2 in a vertical reactor.

$NbCl_5$ (**9**, 133)
Reaction between Nb_2O_5 and refluxing hexachloropropene (4 h). Pale yellow crystals.

$MoCl_5$ (**9**, 133)

Reaction between MoO_3 and refluxing hexachloropropene (15 min).

WCl_5 (**13**, 150)

Photochemical reaction between WCl_6 and tetrachloroethene in a sealed tube. Dark blue–green crystals.

WBr_5 (**20**, 41)

Reaction between W and Br_2 entrained in nitrogen (vertical reactor).

13.2.6 M(VI) halides

WCl_6 (**9**, 135)

Reaction between WO_3 and refluxing hexachloropropene (4 h). Black crystals.

13.3 Amides

Metal dialkylamides are useful starting materials, as the amines generated by proton transfer are, in general, good leaving groups. They are most often prepared from metal halides by ligand metathesis using Group 1 amides, or by elimination of Me_3SiCl in reactions with $R_2N(SiMe_3)$ or $(Me_3Si)NHR$. Dialkyl and bis(trimethylsilyl) amides have been reviewed [4]. A few examples are given below.

$LiNMe_2$ (**21**, 53)

An excess of $HNMe_2$ condensed onto a frozen solution of Bu^nLi in hexane from a calibrated manifold. Mixture allowed to warm slowly while the melt is swirled behind a safety screen. Volatiles removed to leave a white solid.

$Na[NH_2]$ (**2**, 128)

Na added to liquid NH_3 in the presence of an iron catalyst generated from an Fe(II) salt.

$Na[N(SiMe_3)_2]$

NaH reacted with $HN(SiMe_3)_2$ in refluxing toluene overnight. Colourless crystals.

$Ba\{N(SiMe_3)_2\}_2$ [5]

Ba granules treated with $HN(SiMe_3)_2$ in THF. NH_3 bubbled through mixture periodically until all Ba has reacted. Product extracted into pentane. Colourless crystals of THF adduct obtained, but sublimation gives ansolvate.

$M\{N(SiMe_3)_2\}_3$ (M = Sc, Ti, V, Cr, Fe) (**18**, 112)

Reaction between MCl_3 and $Li[N(SiMe_3)_2]$ (M = Sc, Cr, Fe) or between $MCl_3(NMe_3)_2$ and $Li[N(SiMe_3)_2]$ (M = Ti, V). Colourless (M = Sc), blue (M = Ti), brown (M = V), green (M = Cr) or green–black (M = Fe) crystals.

$Ln\{(N(SiMe_3)_2\}_3$ (Ln = La, Ce) [6]
Suspension of $LnCl_3$ in THF treated with $K[N(SiMe_3)_2]$ in toluene. Product sublimed. White (Ln = La) or yellow (Ln = Ce) solids.

$Zr(NMe_2)_4$ [7]
Addition of $LiNMe_2$ to $ZrCl_4$ in THF at 0°C. Isolated and purified by sublimation.

$Mo_2(NMe_2)_6$ (21, 51)
Addition of $MoCl_3$ to $LiNMe_2$ in THF. Purified by sublimation. Yellow crystals.

$Mo(NMe_2)_4$ [8]
$MoCl_5$ added to $LiNMe_2$ in THF/petrol. Product sublimed onto cold finger (solid CO_2) at $40–70°C/10^{-3}$ mm. Purple solid.

$W_2(NMe_2)_6$ (29, 139)
Treatment of WCl_4 with $LiNMe_2$ in Et_2O/THF. Purified by sublimation. Yellow crystals.

$W(NMe_2)_6$ [9]
Addition of $LiNMe_2$ in hexane/THF to $WOCl_4$ in Et_2O at 0°C. Red solid.

$M\{N(SiMePh_2)_2\}_2$ (M = Fe, Co) [10]
Suspension of $FeBr_2$ or $CoCl_2$ in Et_2O treated with a solution of $Li[N((SiMePh_2)_2]$ in Et_2O at 0°C. Linear, two-coordinate complexes.

$M\{N(SiMe_3)_2\}_2$ (M = Ge, Sn, Pb) [11]
Treatment of $MCl_2(Sn, Pb)$ or $GeCl_2$·dioxane in Et_2O with $Li[N(SiMe_3)_2]$ at 0°C. Yellow crystals.

$Bi(NR_2)_3$ (31, 98)
Treatment of $BiCl_3$ with $LiNMe_2$ or $Na[N(SiMe_3)_2]$.

13.4 Alkoxides and related compounds

Alkoxides, aryloxides and siloxides are versatile ancillary ligands, and metal alkoxides find extensive use as precursors to oxide materials. The upsurge in synthetic effort in the area has produced an enormous variety of new alkoxides and several reviews have appeared [12], although the major authoritative work by Bradley et al. is now somewhat dated, and a new review by Bradley and Rothwell will be published soon. General comments on alkoxide synthesis are given below, together with some specific examples.

Alkoxides of the more electropositive metals may be prepared by reacting the metal, a metal hydride or a metal alkyl with an alcohol. For the transition metals, halide metathesis reactions using an appropriate alcohol (often in the presence of a base to scavenge HCl), alkali metal alkoxide or trimethylsilyl ether are commonly employed. An alternative method, which is useful for transition and main-group metals, is the treatment of a dialkylamido compound with an alcohol. In some cases, alkoxo ligand exchange may be achieved by reacting an alkoxide with an alcohol or an ester. Electrochemical syntheses of metal alkoxides have also been developed [13].

LiOR
Treatment of Bu^nLi in hexanes with ROH.
NaOR
Reaction between Na or NaH and ROH.
$Mg(OR)_2$
Reaction between Mg and smaller alcohols, or treatment of MgR_2 with ROH.
$La_2(OCPh_3)_6$ [6]
Ph_3COH in toluene added to solution of $La\{N(SiMe_3)_2\}_3$ in toluene. Colourless crystals.
$M(OMe)_2$ (M = Cr, Mn, Fe, Co, Ni, Cu) and $M(OR)_3$ (M = Ti, Cr, Fe) [14]
Treatment of MCl_n (n = 2, 3) with LiOMe in MeOH. Insoluble products washed with MeOH to remove LiCl.
$Ti(OR)_4$, $Nb(OR)_5$ [15]
Toluene solution of $TiCl_4$ or $NbCl_5$ treated with ROH and NH_3 bubbled through the solution. Product isolated from filtrate after filtering off $[NH_4]Cl$.
$M_2(OR)_6$ (M = Mo, W) [16]
$M_2(NMe_2)_6$ reacted with ROH in hydrocarbon solvent. Amine adducts obtained in some cases.
$Cu(OBu^t)$ [17]
CuCl treated with $LiOBu^t$ in THF, solvent removed and product sublimed from solid residue with IR heating. Yellow crystals.
$Al(OR)_3$ [18]
Reaction between Al and ROH in presence of $HgCl_2$ catalyst. Aryloxides do not require catalyst.
$\{Sn(OBu^t)_2\}_2$ [19]
Hexane solution of $Sn\{N(SiMe_3)_2\}_2$ treated with Bu^tOH. Colourless crystals.
$\{Pb(OBu^t)_2\}_3$ [20]
Hexane solution of $Pb\{N(SiMe_3)_2\}_2$ treated with Bu^tOH. Colourless crystals.

Bi(OBu^t)$_3$ [21]

Reaction between BiCl$_3$ and NaOBu^t in THF. Product extracted into hexane. Colourless solid.

13.5   Oxo compounds

13.5.1   Oxohalides and related compounds

VOCl$_3$ (**9**, 80; **6**, 119; **4**, 80)

Reaction between V$_2$O$_5$ and SOCl$_2$.

NbOCl$_3$ [22]

NbCl$_5$ in CH$_2$Cl$_2$ treated with solution of (Me$_3$Si)$_2$O in CH$_2$Cl$_2$ and the mixture quickly warmed to 80°C and kept at that temperature for 4.5 h.

NbOCl$_3$(MeCN)$_2$ [22]

NbCl$_5$ in MeCN treated with a solution of (Me$_3$Si)$_2$O in MeCN at room temperature. Colourless crystalline solid.

NbOCl$_3$(THF)$_2$ [22]

NbOCl$_3$(MeCN)$_2$ treated with THF.

MoOCl$_4$ (**28**, 325)

From MoO$_3$ and SOCl$_2$. Green crystals, which melt at 101–103°C to give a red–brown liquid.

MoOCl$_3$ (**12**, 190)

Reaction between MoOCl$_4$ and refluxing chlorobenzene. Brown powder, melts to give a black liquid which solidifies to black needles.

MoO$_2$Cl$_2$ [23]

CH$_2$Cl$_2$ solution of (Me$_3$Si)$_2$O added to suspension of MoOCl$_4$ in CH$_2$Cl$_2$.

WOCl$_4$ (**14**, 112)

Solid-state reaction between WO$_3$ and WCl$_6$ in a sealed tube at 200°C. Alternatively, reaction between WO$_3$·H$_2$O and SOCl$_2$ (**29**, 324), or treatment of suspension of WCl$_6$ in CH$_2$Cl$_2$ with a CH$_2$Cl$_2$ solution of (Me$_3$Si)$_2$O [23]. Orange–red needles.

WOCl$_3$ (**14**, 113)

Solid-state reaction between W, WO$_3$ and WCl$_6$ in a sealed tube with a temperature gradient of 450/230°C. Black needles.

WO$_2$Cl$_2$ (**14**, 110)

Solid-state reaction between WO$_3$ and WCl$_6$ in a sealed tube with a temperature gradient of 350/275°C. Yellow plates.

Alternatively, petrol solution of (Me$_3$Si)$_2$O added to suspension of WOCl$_4$ in petrol and mixture heated at 100°C overnight [23].

$WO_2Cl_2(DME)$ [24]

Treatment of a suspension of $WOCl_4$ in CH_2Cl_2 with DME, then with $(Me_3Si)_2O$. Colourless crystals.

$ReOCl_4$ [25]

$ReCl_5$ heated to 250°C in a stream of dry O_2. Product collected in U-tube cooled to −35°C.

ReO_3Cl [25]

Re_3Cl_9 heated in a stream of dry O_2 until it inflames. Product collected in U-tube cooled to −35°C. Colourless liquid m.p. 4.5°C, purified by distillation.

$MeReO_3$ [26]

Re_2O_7 reacted with Me_4Sn in refluxing THF. Sublimed from solid product. Colourless crystals.

FeOCl (**22**, 86)

Solid-state reaction between α-Fe_2O_3 and $FeCl_3$ in a sealed pyrex tube at 370°C for 2–7 days. Red–violet crystals.

13.5.2 Oxometalates

The non-aqueous chemistry of complex polyoxometalates has expanded enormously as a result of advances in modern instrumentation, and synthetic routes to the more important compounds have been published in *Inorg. Synth.* (**27**, 71). Brief details from these and other articles are given below.

$[Bu^n_4N]_3[H_3V_{10}O_{28}]$ (**27**, 83)

Acidified aqueous solution of Na_3VO_4 treated with $[Bu^n_4N]Br$. Precipitate dried and recrystallized from MeCN. Yellow–orange crystals.

$[Bu^n_4N]_2[MoO_4]$

Suspension of MoO_3 in MeCN treated with $[Bu^n_4N]OH$ in MeOH (analogous to synthesis of $[Bu^n_4N]_2[WO_4]$, given below). White solid.

$[Bu^n_4N]_2[Mo_2O_7]$ (**27**, 79)

$[Bu^n_4N]_4[Mo_8O_{26}]$ in acetonitrile treated with $[Bu^n_4N]OH$.

$[Bu^n_4N]_2[Mo_6O_{19}]$ (**27**, 77)

Aqueous solution $Na_2MoO_4 \cdot 2H_2O$ acidified with HCl and then treated with $[Bu^n_4N]Br$. Precipitate filtered, washed, dried and recrystallized from acetone. Yellow crystals.

$[Bu^n_4N]_4[Mo_8O_{26}]$ (**27**, 78)

Aqueous solution $Na_2MoO_4 \cdot 2H_2O$ acidified with HCl and then treated with $[Bu^n_4N]Br$.

$[Bu^n_4N]_2[WO_4]$ [27]

Suspension of $WO_3 \cdot H_2O$ in MeCN reacted with $[Bu^n_4N]OH$ in

MeOH (slight modification of procedure used in reference). White solid.

[Bu^n_4N]$_2$[W_6O_{19}] (**27**, 80)

Mixture of $Na_2WO_4 \cdot 2H_2O$, acetic anhydride and DMF heated to 100°C for 3 h, then a solution of acetic anhydride and conc. HCl in DMF added and the resulting mixture filtered. [Bu^n_4N]Br in MeOH added and precipitate washed with Et_2O and dried.

[Bu^n_4N]$_4$[$W_{10}O_{32}$] (**27**, 81)

Aqueous solution of $Na_2WO_4 \cdot 2H_2O$ acidified quickly with boiling 3 M HCl and the mixture boiled for 1–2 min. [Bu^n_4N]Br added and precipitate collected and dried and recrystallized from DMF or MeCN.

[Bu^n_4N][ReO_4] (**28**, 333)

Slow addition of a hot solution of [Bu^n_4N]Br to one of $KReO_4$. White solid.

13.6 Organoimido complexes

Organoimido (NR) groups are important ligands in early transition-metal chemistry and the synthetic effort in this area has increased dramatically over the last decade or so. The area has been extensively reviewed [28]. The syntheses of a few representative compounds are given below.

{(THF)Mg(NAr)}$_6$ [29]

Treatment of ArH with MgR_2 in THF.

Ti(NPh)Cl$_2$(TMEDA) [30]

Reaction between TiCl$_2$(TMEDA) and PhN=NPh

V(NAr)Cl$_3$ [31]

Reaction between $VOCl_3$ and ArNCO.

CpNb(NR)Cl$_2$ (R = Me, Bu^t) [32]

Reaction between CpNbCl$_4$ and (Me$_3$Si)$_2$NMe or (Me$_3$Si)NHBu^t.

M(NBu^t)$_2$(OSiMe$_3$)$_2$ [33]

Reaction between MO_2Cl_2 and (Me$_3$Si)NHBu^t in refluxing hexane.

Cr(NAr)$_2$Cl$_2$ [34]

Treatment of Cr(NBu^t)$_2$Cl$_2$ with ArNH$_2$.

Mo(NC$_6$H$_4$Me-4)Cl$_4$(THF) [35]

Reaction between MoCl$_4$(THF)$_2$ and 4-MeC$_6$H$_4$N$_3$.

Mo(NBu^t)$_2$Cl$_2$(DME) [36]

Treatment of Na_2MoO_4 with Et$_3$N, Bu^{t}NH$_2$ and an excess of Me$_3$SiCl in DME.

W(NR)Cl$_4$ (**24**, 194)

Reaction between WOCl$_4$ and RNCO in refluxing benzene or toluene.

Re(NR)Cl$_4$ (**24**, 194)

Reaction between ReOCl$_4$ and RNCO in refluxing benzene or toluene.

Os(NBu^t)$_3$O [37]

Reaction between OsO$_4$ and Bu$^n_3$P$=$NBu^t.

13.7   Carbonyls

Most of the simpler transition-metal carbonyls are commercially available, but the book by King gives details of syntheses using autoclave techniques [38]. Some representative syntheses of their derivatives are included below.

Na[Nb(CO)$_6$] (**28**, 193)

NbCl$_5$ added to stirred mixture of Mg and Zn powders in anhydrous pyridine under an atmosphere of CO. Resulting solution treated with aqueous NaOH and extracted with diethyl ether. Solvent removed and residue treated with THF. Yellow–orange solid (THF solvate).

Cr(η^6-arene)(CO)$_3$ (**28**, 136)

Reaction between arene, THF and Cr(CO)$_6$ in refluxing dibutyl ether.

M(CO)$_3$(EtCN)$_3$ (M = Cr, Mo, W) (**28**, 29)

M(CO)$_6$ in refluxing EtCN. Mo reaction much faster. Light yellow (Mo, W) or yellow (Cr) crystalline solids.

MoBr$_2$(CO)$_4$ (**28**, 145)

Mo(CO)$_6$ in CH$_2$Cl$_2$ treated with Br$_2$.

Na[Mn(CO)$_5$] (**28**, 199)

Na/Hg amalgam reduction of Mn$_2$(CO)$_{10}$ in THF. Solution used for further reactions.

Mn(CO)$_5$Cl (**28**, 155)

Mn$_2$(CO)$_{10}$ in CCl$_4$ at 0°C treated with solution of Cl$_2$ in CCl$_4$. Yellow solid.

Mn(CO)$_5$Br (**28**, 156)

Solution of Mn$_2$(CO)$_{10}$ in CS$_2$ treated with a solution of Br$_2$ in CS$_2$. Yellow–orange solid.

Mn(CO)$_5$I (**28**, 157)

Treatment of Na[Mn(CO)$_5$] with I$_2$ in THF. Red–orange crystals.

$Re_2(CO)_{10}$ [39]

Low pressure synthesis from $[NH_4][ReO_4]$ and $(Me_2CHCH_2)_2AlH$ in toluene at 70–80°C under CO atmosphere.

$K[HFe(CO)_4]$ (29, 152)

$Fe(CO)_5$ added to ethanolic KOH. Pale pink solution used for further reactions.

$Na_2[Fe(CO)_4]$ (28, 204)

Solution of $Fe(CO)_5$ in THF added to sodium naphthalenide in THF. White pyrophoric solid.

$Na_2[Fe_2(CO)_8]$ (28, 204)

As above, but with $Fe:Na[C_{10}H_8]$ ratio of 1:1. Orange pyrophoric solid.

$Fe(CO)_3(PPh_3)_2$ (29, 153)

PPh_3 added to a solution of $K[HFe(CO)_4]$ and the mixture refluxed for 24 h. Yellow powder.

$Ru_3(CO)_{12}$ (28, 216)

High-pressure reaction between $RuCl_3 \cdot xH_2O$ and CO in MeOH. Air-stable orange solid.

$Ru_4H_4(CO)_{12}$ (28, 219)

Reaction between $Ru_3(CO)_{12}$ and H_2 in refluxing octane or in cyclohexane under pressure. Air-stable yellow powder.

$Ru_3(CO)_{11}(PMe_2Ph)$ (28, 223)

$Na[Ph_2CO]$ solution added to a mixture of $Ru_3(CO)_{12}$ and PMe_2Ph in THF. Red–orange crystals.

$Os_3(CO)_{12}$ (28, 230)

High-pressure reaction between OsO_4 and CO in EtOH. Yellow crystals.

$Os_3(CO)_{11}(MeCN)$ (28, 232)

MeCN solution of $Os_3(CO)_{12}$ treated with freshly sublimed Me_3NO. Air-stable yellow crystals.

$Os_4H_4(CO)_{12}$ (28, 240)

High-pressure reaction between $Os_3(CO)_{12}$ and H_2 at 140°C. Pale yellow solid.

$[PPN][M(CO)_4]$ (M = Rh, Ir) (28, 211)

KOH added to $MCl_3 \cdot xH_2O$ in DMSO under CO. Addition of $[(PPh_3)_2N]^+Cl$ precipitates product. White crystalline solids.

13.8 Cyclopentadienyl complexes

$CpTiCl_3$ (ref. 38, p. 78.)

Mixture of $TiCl_4$ and Cp_2TiCl_2 heated in refluxing xylene for 2.5 h. Yellow crystals.

Cp_2TiCl_2 (ref. 38, p. 75.)

Addition of CpNa(DME) to solution of $TiCl_4$ in aromatic hydrocarbon. Red crystals.

Cp_2TiCl (**28**, 261)

$TiCl_3$ treated with TlCp in THF under reflux. Green–yellow solid, very air-sensitive.

$Cp_2M(CO)_2$ (M = Ti, Zr, Hf) (**28**, 248)

Reduction of Cp_2MCl_2 with $Mg/HgCl_2$ in THF under CO atmosphere. Note: Hf reduction requires Mg powder, whereas turnings are used for Ti and Zr reductions. Red (M = Ti), black (M = Zr) or purple (M = Hf) air-sensitive solids.

Cp_2ZrH_2 (**28**, 257)

Cp_2ZrCl_2 reacted with H_2O and aniline to give $\{Cp_2ZrCl\}_2O$, which is isolated and then treated with standardized solution of $Li[AlH_4]$ in THF. Colourless microcrystalline solid.

Cp_2ZrHCl (**28**, 257)

Cp_2ZrCl_2 treated with $Li[AlH(OBu^t)_3]$ in THF. Colourless crystalline solid.

Cp_2V (**28**, 263)

$[V_2Cl_6(THF)_6][Zn_2Cl_6]$ added to a cooled solution of NaCp in THF. Mixture allowed to warm, and then refluxed. Product sublimed. Violet air-sensitive crystals.

Cp_2VCl (**28**, 262)

VCl_3 treated with TlCp in THF under reflux. Blue–black solid, extremely air-sensitive.

$CpNbCl_4$ [40]

Solution of $NbCl_5$ in CH_2Cl_2 treated with $Bu^n_3Sn(\sigma\text{-}C_5H_5)$. Maroon crystals.

$Cp_2Mo_2(CO)_6$ (**28**, 150)

Treatment of solid $Mo(CO)_3(NCR)_3$ with excess cyclopentadiene under reflux. Red crystals.

$Cp_2M_2(CO)_6$ (M = Mo, W) [41]

$Na[CpM(CO)_3]$ prepared *in situ* from $M(CO)_6$ and NaCp (from NaH and CpH) in refluxing THF (M = Mo) or diglyme (M = W). Resulting solution oxidized with a solution of Fe(III). This is a slight modification of the procedure described in the reference. Red or red–brown crystals.

$Cp_2M_2(CO)_4$ (**28**, 152)

Thermolysis of $Cp_2M_2(CO)_6$ in diglyme. Longer reaction times needed for tungsten compound.

Cp_2MoCl_2 (**29**, 208)

Cp_2MoH_2 reacted with $CHCl_3$ under reflux. Green solid.

Alternatively from reaction between $Mo_2(O_2CCH_3)_4$, NaCp and PPh_3, followed by addition of HCl [42].

Cp_2MoH_2 (**29**, 205)

Slow addition of THF to a cooled slurry of $MoCl_5$, $NaCp \cdot DME$ and $Na[BH_4]$ in hexanes. Mixture refluxed, then treated with 6 M HCl. Yellow solid, purified by sublimation. (Procedure can take 5–6 days to complete.)

Cp_2MoO (**29**, 209)

Cp_2MoCl_2 treated with aqueous NaOH. Product purified by chromatography on specially treated silica. Emerald green solid.

$CpRe(CO)_3$ (**29**, 21)

Reaction between $Re_2(CO)_{10}$ and dicyclopentadiene under reflux. Off-white solid.

Cp^*ReO_3 ($Cp^* = \eta^5-C_5Me_5$) [43]

Benzene solution of $Cp^*Re(CO)_3$ treated with aqueous H_2O_2 under reflux.

$\{CpFe(CO)_2\}_2$ (ref. 38, p. 114.)

Mixture of $Fe(CO)_5$ and dicyclopentadiene refluxed for 16 h. Red–purple crystals.

$Na[CpFe(CO)_2]$ (**28**, 208)

Reduction of $Cp_2Fe_2(CO)_4$ with Na/Hg amalgam in THF. Reaction monitored by IR spectroscopy.

$[CpFe(CO)(\eta^2-C_4H_8)][BF_4]$ (**28**, 210)

$Na[CpFe(CO)_3]$ treated with 3-chloro-2-methyl-1-propene (isobutenyl chloride) to give $CpFe(CO)_2\{\eta^1-CH(Me)=CH_2\}$, which is then treated with $H[BF_4]$. Yellow crystals.

$\{CpRu(CO)_2\}_2$ (**28**, 189)

$Ru_3(CO)_{12}$ treated with cyclopentadiene in refluxing heptane to give $CpRuH(CO)_2$, which is then heated in refluxing heptane in air. Purified by chromatography on alumina. Orange crystals.

$\{Cp^*RuCl_2\}_2$ ($Cp^* = \eta^5-C_5Me_5$) (**29**, 226)

$RuCl_3 \cdot xH_2O$ reacted with C_5Me_5H in refluxing MeOH.

$Cp^*Co(CO)_2$ ($Cp^* = \eta^5-C_5Me_5$) (**28**, 273)

Mixture of $Co_2(CO)_8$, C_5Me_5H and 1,3-cyclohexadiene heated in refluxing CH_2Cl_2. Red crystals.

$\{Cp^*MCl_2\}_2$ ($Cp^* = \eta^5-C_5Me_5$; M = Rh, Ir) (**29**, 229)

Reaction between $MCl_3 \cdot xH_2O$ and C_5Me_5H in refluxing MeOH. Red (M = Rh) or orange (M = Ir) crystals.

$CpTl$ (**24**, 97)

Aqueous solution of Tl_2SO_4 and NaOH stirred with freshly distilled cyclopentadiene for 12 h. Purified by sublimation at 90–100°C/10^{-3} mm. Pale yellow solid.

13.9   Miscellaneous metal complexes

13.9.1   β-Diketonates and carboxylates

$VO(O_2CCH_3)_2$ (13, 181)
Reaction between V_2O_5 and $(CH_3CO)_2O$ under reflux.

$VO(acac)_2$ (5, 113)
Mixture of V_2O_5 and EtOH in aqueous H_2SO_4 heated for 30 min to give a blue solution. AcacH added, followed by Na_2CO_3 to neutralize the acid. Blue or blue–green crystals.

$TiCl_2(acac)_2$ (19, 145)
Solution of 2,4-pentanedione (acacH) in CH_2Cl_2 added to a solution of $TiCl_4$ in CH_2Cl_2. Orange–red solid.

$MoO_2(acac)_2$ (29, 130)
$[NH_4]_6[Mo_7O_{24}]$ dissolved in aqueous ammonia and treated with acacH and then conc. HNO_3. Pale yellow crystals.

$Cr_2(O_2CCH_3)_4$ (8, 125)
Acidified solution of hydrated $CrCl_3$ reduced with amalgamated zinc (Jones reductor) and added to solution of $Na[O_2CCH_3]$. Red crystalline hydrate heated *in vacuo* to give brown anhydrous product.

$Mo_2(O_2CCH_3)_4$ (13, 81)
Reaction between $Mo(CO)_6$ and $CH_3CO_2H/(CH_3CO)_2O$ at elevated temperatures in 1,2-dichlorobenzene.
$Mo_2(O_2CR)_2$ can also be prepared from $Mo_2Br_4(PBu^n_3)_4$ and the appropriate acid in benzene (19, 132).

$W_2(O_2CBu^t)_4$ (26, 223)
Mixture of WCl_4 and $Na[O_2CBu^t]$ in THF reduced with Na/Hg amalgam. Yellow crystals.

13.9.2   Thio and alkylthiolate complexes

$M(S)Cl_3$, $M(S)Cl_3(MeCN)_2$ (M = Nb, Ta)
$Mo(S)Cl_3$
$W(S)Cl_4$, $W(S)_2Cl_2$
$M(O)(S)Cl_2$ (M = Mo, W)
The above compounds can be prepared by low temperature reactions between $(Me_3Si)_2S$ and the appropriate metal halide or oxohalide [44].

$[PPh_4]_2[M(SPh)_4]$ (M = Mn, Fe, Co, Zn, Cd) (21, 23)
Dithiocarbonato complexes of M(II) treated with K[SPh]. Products precipitated by addition of $[PPh_4]Cl$.

13.9.3 Pd(0) and Pt(0) phosphine complexes

Pd(PR$_3$)$_2$ (**19**, 101)
Pd(η^3-C$_3$H$_5$)(η^5-C$_5$H$_5$) in toluene treated with a solution of a bulky phosphine.
Pd(PPh$_3$)$_4$ (**13**, 121)
Solution of PdCl$_2$ and PPh$_3$ in hot DMSO reduced with hydrazine hydrate. Yellow crystalline product filtered from cooled solution.
Pt(PR$_3$)$_2$ (**19**, 104)
PtCl$_2$(PR$_3$)$_2$ prepared from K$_2$[PtCl$_4$] and the bulky phosphine, then dissolved in THF and reduced with Na/Hg amalgam or Na[C$_{10}$H$_8$].
Pt(PPr$^i_3$)$_3$ (**19**, 108)
PtCl$_2$(PPr$^i_3$)$_2$ prepared from K$_2$[PtCl$_4$] and PPr$^i_3$, then dissolved in THF and reduced with Na/Hg amalgam in the presence of PPr$^i_3$.
Pt(PEt$_3$)$_4$ (**19**, 110)
Aqueous solution of K$_2$[PtCl$_4$] added to a mixture of KOH and PEt$_3$ in EtOH/H$_2$O (30:1) to give a reddish oil, which is then treated with PEt$_3$. Colourless crystals.

13.9.4 Complexes with weakly bound ligands

[M(NO)$_2$(MeCN)$_4$][BF$_4$] (M = Mo, W) (**28**, 65)
Reaction between M(CO)$_6$ and [NO][BF$_4$] in MeCN. Dark green crystals.
[Pd(MeCN)$_4$][BF$_4$]$_2$ (**28**, 63)
Reaction between Pd sponge and [NO][BF$_4$] in MeCN. Pale yellow crystals.
[Cu(MeCN)$_4$][PF$_6$] (**28**, 68)
Suspension of Cu$_2$O in MeCN treated with 60–65% solution of HBF$_4$. White microcrystalline solid.
M(CF$_3$SO$_3$)$_n$ (**28**, 72)
MCl$_n$ treated with aqueous CF$_3$SO$_3$H. Product precipitated with Et$_2$O.

13.10 Ligands and reagents

13.10.1 Magnesium alkyls

Mg(CH$_2$CMe$_3$)Cl (**26**, 46)
Mg turnings in Et$_2$O treated with neopentyl chloride and mixture heated to reflux. 1,2-Dibromoethane added to initiate reaction if necessary.

Mg(CH$_2$SiMe$_3$)$_2$ (**19**, 262)

Generation of Mg(CH$_2$SiMe$_3$)Cl from Me$_3$SiCH$_2$Cl and Mg in Et$_2$O, then addition of 1,4-dioxane. Isolated as white solid. Standard solutions in Et$_2$O can be stored at 0–5°C. Procedure generally applicable to MgR$_2$ compounds.

13.10.2 Phosphorus compounds

R$_2$PPh (R = Et, Bu^n, C$_6$H$_{11}$, CH$_2$Ph) (**18**, 169)

Treatment of RMgX with PhPCl$_2$ in Et$_2$O.

PMe$_3$ (**28**, 305)

From P(OPh)$_3$ and MeMgBr in Bu$^n_2$O. Colourless, pyrophoric liquid with unpleasant odour; b.p. 39–40°C.

Bu^{t}PCl$_2$ and Bu$^t_2$PCl (**14**, 4)

From Bu^{t}MgCl and PCl$_3$. Mixtures of these compounds separated by fractional distillation (Bu^{t}PCl$_2$ m.p. *ca.* 50°C, b.p. *ca.* 60°C/12 mm; Bu$^t_2$PCl m.p. 2–3°C, b.p. *ca.* 70°C/13 mm). Useful for the preparation of bulky phosphines.

P(SiMe$_3$)$_3$ (**27**, 243)

Na/K alloy added to white phosphorus in DME at 50°C, then Me$_3$SiCl added to the resulting suspension. Purified by distillation, b.p. 30–35°C/10^{-3} mm, m.p. *ca.* 25°C.

Ar*PH$_2$ (Ar* = 2,4,6-Bu$^t_3$C$_6$H$_2$) (**27**, 236)

Ar*Li (generated from Ar*Br and Bu^{n}Li) added to PCl$_3$ in THF. Ar*PCl$_2$ is then reacted with Li[AlH$_4$] in THF; m.p. 144–146°C.

13.10.3 Silicon and tin reagents

Me$_3$SiBr (**26**, 4)

From PPh$_3$Br$_2$ and (Me$_3$Si)$_2$O in refluxing 1,2-dichlorobenzene with a catalytic amount of Zn.

Me$_3$SiOMe (**26**, 44)

Reaction between (Me$_3$Si)$_2$NH and MeOH in the presence of a small amount of Me$_3$SiCl; b.p. 57°C.

Me$_3$SiNCO (**25**, 48)

Mixture of (Me$_3$Si)$_3$N and AlCl$_3$ treated with SOCl$_2$, then heated to 70°C for 24 h. Yellow liquid, b.p. 102–104°C.

R$_n$Si(NCO)$_{4-n}$ (**24**, 99)

R$_n$SiCl$_{4-n}$ in liquid SO$_2$ treated with K[NCO].

(Me$_3$Si)$_2$S (**29**, 30)

Na$_2$S (prepared *in situ* from Na and sulphur) in THF treated with Me$_3$SiCl. Colourless, foul-smelling liquid; b.p. 162°C/760 mm, 91–95°C/100 mm.

Bu^n_3SnSPh (**25**, 114)

$(Bu^n_3Sn)_2O$ treated with PhSH. Water formed in reaction removed by distillation. Viscous liquid, b.p. 172–174°C/2.5 mm.

13.10.4 Miscellaneous compounds

Pyrazolylborates (**12**, 99)

$K[BH_4]$ and pyrazole heated to melting point. Product isolated by pouring the molten product into toluene. These ligands find extensive use in coordination chemistry and substituted derivatives have been described in the literature [45].

C_5Me_5H (**29**, 193)

Reaction between 3-pentanone and acetaldehyde to give 2,3,5,6-tetrahydro-2,3,5,6-tetramethyl-γ-pyrone, which is then converted to 2,3,4,5,-tetramethylcyclopent-2-enone by treatment with formic acid and conc. H_2SO_4. The ketone is treated with LiMe and then acidified to give C_5Me_5H.

$[(Ph_3P)_2N]Cl$ (**15**, 84)

Reaction between Ph_3PCl_2 and hydroxylamine hydrochloride in the presence of PPh_3 in refluxing 1,1,2,2-tetrachloroethane solvent.

References

1. Edema, J.J.H., Duchateau, R., Gambarotta, S., et al. (1991) Inorg. Chem., **30**, 154.
2. Chisholm, M.H., Eichorn, B.W., Folting, K., et al. (1987) Inorg. Chem., **26**, 3183.
3. Schrock, R.R., Sturgeoff, L.G. and Sharp, P.R. (1983) Inorg. Chem., **22**, 2801.
4. Bradley, D.C. and Chisholm, M.H. (1976) Acc. Chem. Res., **9**, 273; Eller, P.G., Bradley, D.C., Hursthouse, M.B. and Meek, D.W. (1977) Coord. Chem. Rev., **24**, 1; Harris, D.H., Lappert, M.F., Power, P.P., et al. (1980) Metal and Metalloid Amides, Ellis Horwood, Chichester.
5. Caulton, K.G., Huffman, J.C., Streib, W.E. and Vaarstra, B.A. (1991) Inorg. Chem., **30**, 121.
6. Evans, W.J., Golden, R.E. and Ziller, J.W. (1991) Inorg. Chem., **30**, 4963.
7. Chisholm, M.H., Hammond, C.E. and Huffman, J.C. (1988) Polyhedron, **7**, 2515.
8. Bradley, D.C. and Chisholm, M.H. (1960) J. Chem. Soc., 3857.
9. Chisholm, M.H., Cotton, F.A., Extine, M.W. and Stults, B.R. (1976) J. Am. Chem. Soc., **98**, 4477.
10. Barlett, R.A. and Power, P.P. (1987) J. Am. Chem. Soc., **109**, 7563.
11. Harris, D.H. and Lappert, M.F. (1974) J. Chem. Soc., Chem. Commun., 895.
12. Bradley, D.C., Mehrotra, R.C. and Gaur, D.P. (1978) Metal Alkoxides, Academic Press, New York; Mehrotra, R.C. (1983) Adv. Inorg. Chem. Radiochem., **26**, 269; Chisholm, M.H. (1983) Polyhedron, **2**, 681; Mehrotra, R.C., Singh, A. and Sogani, S. (1994) Chem. Rev., **94**, 1643; Herrman, W.A., Huber, N.W. and Runte, O. (1995) Angew. Chem., Int. Ed. Eng., **34**, 2187.
13. Shreider, V.A., Turevskaya, E.P., Koslova, N.I. and Turova, N.Y. (1981) Inorg. Chim. Acta, **53**, L73.
14. Adams, R.W., Bishop, E., Martin, R.L. and Winter, G. (1966) Aust. J. Chem., **19**, 207.
15. Bradley, D.C., Gaze, R. and Wardlaw, W. (1955) J. Chem. Soc., 721; Bradley, D.C., Chakravarti, B.N. and Wardlaw, W. (1956) J. Chem. Soc., 2381.

16. Chisholm, M.H. and Reichert, W. (1974) *J. Am. Chem. Soc.*, **96**, 1249; Chisholm, M.H. and Extine, M. (1975) *J. Am. Chem. Soc.*, **97**, 5625.
17. Tsuda, T., Hashimoto, T. and Saegusa, T. (1972) *J. Am. Chem. Soc.*, **94**, 658; Lemmen, T.H., Goeden, G.V., Huffman, J.C. and Caulton, K.G. (1990) *Inorg. Chem.*, **29**, 3680.
18. Mehrotra, R.C. and Rai, A.K. (1991) *Polyhedron*, **10**, 1967.
19. Veith, M. and Töllner, F. (1983) *J. Organomet. Chem.*, **246**, 219; Fjeldberg, T., Hitchcock, P.B., Lappert, M.F., *et al.* (1985) *J. Chem. Soc., Chem. Commun.*, 939.
20. Goel, S.C., Chiang, M.Y. and Buhro, W. (1990) *Inorg. Chem.*, **29**, 4640.
21. Evans, W.J., Hain, Jr. J.H. and Ziller, J.W. (1989) *J. Chem. Soc., Chem. Commun.*, 1628.
22. Gibson, V.C., Kee, T.P. and Shaw, A. (1988) *Polyhedron*, **7**, 2217.
23. Gibson, V.C., Kee, T.P. and Shaw, A. (1988) *Polyhedron*, **7**, 579.
24. Dreisch, K., Andersson, C. and Stålhandske, C. (1991) *Polyhedron*, **10**, 2417.
25. Edwards, P.G., Wilkinson, G., Hursthouse, M.B. and Malik, K.M.A. (1980) *J. Chem. Soc., Dalton Trans.*, 2467.
26. Herrman, W.A., Kuchler, J.G., Felixberger, J.K., *et al.* (1988) *Angew. Chem. Int. Ed. Eng.*, **27**, 394.
27. Che, T.M., Day, V.W., Francesconi, L.C., *et al.* (1985) *Inorg. Chem.*, **24**, 4055.
28. Nugent, W.A. and Haymore, B.L. (1980) *Coord. Chem. Rev.*, **31**, 123; Wigley, D.E. (1994) *Prog. Inorg. Chem.*, **42**, 239.
29. Grigsby, W.J., Hascall, T., Ellison, J.J., *et al.* (1996) *Inorg. Chem.*, **35**, 3254.
30. Duchateau, R., Williams, A.J., Gambarotta, S. and Chiang, M.Y. (1991) *Inorg. Chem.*, 1991, **30**, 4863.
31. Maatta, E.A. (1984) *Inorg. Chem.*, **23**, 2560.
32. Williams, D.N., Mitchell, J.P., Poole, A.D., *et al.* (1992) *J. Chem. Soc., Dalton Trans.*, 739.
33. Nugent, W.A. and Harlow, R.L. (1980) *Inorg. Chem.*, **19**, 777.
34. Coles, M.P., Dalby, C.L., Gibson, V.C., *et al.* (1995) *Polyhedron*, **14**, 2455.
35. Chou, C.Y., Huffman, J.C. and Maatta, E.A. (1984) *J. Chem. Soc., Chem. Commun.*, 1184.
36. Dyer, P.W., Gibson, V.C., Howard, J.A.K., *et al.* (1992) *J. Chem. Soc., Chem. Commun.*, 1666.
37. Chong, A.O., Oshima, K. and Sharpless, K.B. (1977) *J. Am. Chem. Soc.*, **99**, 3420.
38. King, R.B. (1965) *Organometallic Syntheses, Vol. 1, Transition Metal Compounds*, Academic Press, New York.
39. Top, S., Morel, P., Pankowski, M. and Jaouen, G. (1996) *J. Chem. Soc., Dalton Trans.*, 3611.
40. Bunker, M.J., DeCian, A., Green, M.L.H., *et al.* (1980) *J. Chem. Soc., Dalton Trans.*, 2155.
41. Birdwhistel, R., Hackett, P. and Manning, A.R. (1978) *J. Organomet. Chem.*, **157**, 239.
42. Green, M.L.H., Poveda, M.L., Bashkin, J. and Prout, K. (1982) *J. Chem. Soc., Chem. Commun.*, 30; Bashkin, J., Green, M.L.H., Poveda, M.L. and Prout, K. (1982) *J. Chem. Soc., Dalton Trans.*, 2485.
43. Herrmann, W.A., Voss, E. and Flöel, M. (1985) *J. Organomet. Chem.*, **297**, C5; Herrmann, W.A. and Okuda, J. (1987) *J. Mol. Catal.*, **41**, 109.
44. Gibson, V.C., Shaw, A. and Williams, D.N. (1989) *Polyhedron*, **8**, 549.
45. Trofimenko, S. (1986) *Prog. Inorg. Chem.*, **34**, 115; Trofimenko, S., Calabrese, J.C. and Thompson, J.S. (1987) *Inorg. Chem.*, **26**, 1507.

Appendix A   Health and Safety Information

The techniques described earlier in this book enable hazardous, air-sensitive materials to be handled safely on a routine basis, provided manipulations are carefully planned and executed to avoid explosions and fires. However, it is important not to overlook the hazards presented by what might be regarded as more routine reagents, e.g. flammable or toxic solvents, strong acids and bases, and potentially unstable mixtures of oxidizing and reducing agents. Data sheets from suppliers and the references given in chapter 2 and at the end of this appendix should be consulted for detailed hazard information on specific chemicals, but for general reference, common chemical hazards are summarized here in tabular form. In addition, the resistance to chemicals of various glove materials is compared, and methods are summarized for the removal of toxic cations from liquid residues. Solvent flammability data can be found in Appendix C.

A.1 Incompatible chemicals

In Tables A.1 and A.2, substances in the left hand column should be stored and handled so they cannot possibly contact substances in the second column under uncontrolled conditions, when violent reactions may occur (Table A.1) or toxic materials (Table A.2, third column) may be produced.

A.2 Explosion hazards

Examples of shock- or friction-sensitive compounds, and of potentially explosive mixtures are given in Table A.3 and Table A.4, respectively.

A.3 Water-reactive chemicals

Many of the reagents encountered in synthetic inorganic and metalorganic chemistry react sufficiently violently with water to warrant special consideration in a risk assessment, although most can be handled safely using

Table A.1 Incompatible chemicals (reactive hazards)

Acetic acid	Chromic acid, nitric acid, peroxides, permanganates
Acetic anhydride	Hydroxyl-containing compounds, perchloric acid
Acetone	Conc. nitric and sulphuric acid mixtures, hydrogen peroxide
Acetylene	Chlorine, bromine, fluorine, copper, silver, mercury
Aluminium powder	Halogenated or oxygenated solvents
Ammonia (anhydrous)	Mercury, chlorine, bromine, iodine, calcium hypochlorite, hydrogen fluoride
Ammonium nitrate	Acids, metal powders, flammable liquids, chlorates, nitrites, sulphur, finely divided organics or combustibles
Bromine	Ammonia, acetylene, butadiene, gaseous hydrocarbons, sodium carbide, benzene, finely divided metals
Carbon, activated	Calcium hypochlorite, other oxidants
Chlorates	Ammonium salts, acids, metal powders, phosphorus, sulphur, finely divided organics or combustibles
Chromic acid and CrO_3	Acetic acid, naphthalene, camphor, glycerol, alcohol, other flammable liquids
Chlorine	Ammonia, acetylene, butadiene, gaseous hydrocarbons, hydrogen, sodium carbide, benzene, finely divided metals
Copper	Acetylene, hydrogen peroxide
Fluorine	Isolate from all other chemicals
Hydrazine	Hydrogen peroxide, nitric acid, other oxidants, heavy metals
Hydrocarbons	Fluorine, chlorine, bromine, chromic acid, conc. nitric acid, peroxides
Hydrogen peroxide	Copper, chromium, iron, most metals or their salts, flammable liquids, combustible materials, aniline, nitromethane
Hydrogen sulphide	Fuming nitric acid, oxidizing gases
Iodine	Acetylene, ammonia (anhydrous or aqueous)
Mercury	Acetylene, nitric acid–ethanol mixtures, ammonia
Metals of Groups 1 and 2 (e.g. Li, Na, K, Mg, Ca)	Water, chlorinated hydrocarbons, carbon dioxide, halogens (use dry sand in case of fire)
Nitric acid (conc.)	Acetic acid, aniline, chromic acid, hydrogen sulphide, flammable liquids, flammable gases, copper, brass, heavy metals
Nitroalkanes (lower mol. wt.)	Inorganic bases, amines, halogens, 13X molecular sieve
Oxygen	Oils, grease, hydrogen, flammable liquids/solids/gases
Perchloric acid	Acetic anhydride, bismuth and its alloys, alcohol, paper, wood, grease, oils
Peroxides, organic	Acids (organic or mineral), avoid friction, store cold
Phosphorus, white	Air, oxygen
Potassium chlorate	Acids (see also chlorates)
Potassium perchlorate	Acids (see also perchloric acid)
Potassium permanganate	Glycerol, ethylene glycol, benzaldehyde, sulphuric acid
Silver	Acetylene, oxalic acid, tartaric acid, nitric acid–ethanol mixtures, ammonium compounds
Sodium nitrite	Ammonium nitrate and other ammonium salts
Sodium peroxide	Any oxidizable substance
Sulphuric acid	Chlorates, perchlorates, permanganates
Thiocyanates	Metal nitrates, nitrites, oxidants
Trifluoromethanesulphonic acid	Perchlorate salts

Table A.2 Incompatible chemicals (toxic hazards)

Arsenical materials	Any reducing agent	Arsine
Azides	Acids	Hydrogen azide
Cyanides	Acids	Hydrogen cyanide
Hypochlorites	Acids	Chlorine or hypochlorous acid
Nitrates	Sulphuric acid	Nitrogen dioxide
Nitric acid	Copper, brass, heavy metals	Nitrogen dioxide (nitrous fumes)
Nitrites	Acids	Nitrous fumes
Phosphorus	Strong alkalis or reducing agents	Phosphine
Sulphides	Acids	Hydrogen sulphide
Tellurides	Reducing agents	Hydrogen telluride

Table A.3 Shock-sensitive materials

Compound	Comments
Acetylenic compounds	especially polyacetylenes, haloacetylenes, heavy metal acetylides (Cu, Ag, Hg)
Alkyl and acyl nitrates	especially polyol nitrates, e.g. nitrocellulose and nitrolglycerine
Alkyl and acyl nitrites	
Alkyl perchlorates	
Ammine metal oxosalts	metal complexes with coordinated ammonia, hydrazine, or similar nitrogenous bases with oxidizing oxoanions such as ClO_4^-
Azides	metal, non-metal and organic azides
Chlorite salts of metals	e.g. $AgClO_2$ and $Hg(ClO_2)_2$
Diazocompounds	e.g. CH_2N_2
Diazonium salts	when dry
Fulminates	e.g. AgCNO
Hydrogen peroxide	solutions with concentrations >30% form explosive mixtures with organic compounds and decompose violently in the presence of trace amounts of transition metals.
Perchlorate salts	most metal, non-metal and amine perchlorates can be detonated, and may react violently with oxidizable materials
Peroxides	organic and transition-metal peroxides
Picrates	especially salts of heavy metals (Ni, Cu, Zn, Hg, Pb)
Picric acid	not as sensitive to shock or friction as its metal salts, and is safe as a paste wetted with water.

Schlenk and dry-box techniques. Some general classes of such compounds are listed in Table A.5.

A.4 Pyrophoric chemicals

General classes of compounds which are likely to ignite spontaneously in air (by reaction with oxygen and/or moisture) are given in Table A.6. Again, these compounds can be handled safely using inert-atmosphere techniques, but the disposal of residues should be planned carefully.

Table A.4 Potentially explosive mixtures of common reagents

Acetone	+	chloroform in the presence of base
Acetylene	+	copper, silver, mercury, or their salts
Ammonia (inc. aqueous solutions)	+	chlorine, bromine, iodine
Carbon disulphide	+	sodium azide
Chlorine	+	an alcohol
Chlorocarbon	+	powdered Al or Mg
Decolourizing carbon	+	an oxidizing agent
Diethyl ether	+	chlorine
Dimethyl sulphoxide	+	an acyl halide, $SOCl_2$, or $POCl_3$
Dimethyl sulphoxide	+	CrO_3
Ethanol	+	calcium hypochlorite
Ethanol	+	silver nitrate
Nitric acid	+	acetic anhydride or acetic acid
Picric acid (trinitrophenol)	+	a heavy metal salt (e.g. of Pb, Hg, or Ag)
Silver oxide	+	ammonia + ethanol
Sodium	+	a chlorinated solvent
Sodium hypochlorite	+	an amine

Table A.5 Common chemicals which react violently with water

Alkali metal hydrides	e.g. NaH, KH
Alkali metal amides	e.g. $NaNH_2$, $LiNMe_2$
Main group alkyls	e.g. LiR, MgR_2, AlR_3, ZnR_2
Metals of Groups 1 and 2	e.g. Na, K, Ca
Non-metal halides	e.g. BCl_3, BF_3, PCl_3, $SiCl_4$
Inorganic acid halides	e.g. $POCl_3$, $SOCl_2$
Anhydrous covalent metal halides	e.g. $AlCl_3$, $TiCl_4$, $ZrCl_4$
Phosphorus(V) oxide)	P_4O_{10}
Organic acid halides and low molecular weight anhydrides	

Table A.6 Pyrophoric chemicals

Metal alkyls and aryls	e.g. RMgX, RLi, RNa, R_3Al, R_2Zn
Metal carbonyls	e.g. $Ni(CO)_4$, $Fe(CO)_5$, $Co_2(CO)_8$
Alkali metals	e.g. Na, K
Metal powders	e.g. Al, Co, Fe, Mg, Mn, Pd, Pt, Ti, Sn, Zn, Zr
Metal hydrides	e.g. NaH, KH, $LiAlH_4$
Non-metal hydrides	e.g. B_2H_6 and higher boranes, PH_3, AsH_3
Non-metal alkyls	e.g. R_3B, R_3P, R_3As
Phosphorus (white)	

Table A.7 Chemical resistance of glove materials[a]

Chemical	Natural rubber	light-weight	heavy-weight	Nitrile	heavy-weight	Neoprene	Vinyl
Acetaldehyde	G	–	–	E	–	G	G
Acetic acid	E	4	43	E	118	E	E
Acetone	G	1	5	G	9	G	F
Acetonitrile	–	1	6	–	15	–	–
Ammonium hydroxide (0.88)	G	1	12	E	276	E	E
Aniline	F	–	–	E	–	G	G
Anionic detergent, 5%	–	480	480	–	480	–	–
Benzaldehyde	F	–	–	E	–	F	G
Benzene[b]	P	NR	5	G	25	F	F
Benzyl chloride[b]	F	–	–	G	–	P	P
Biological detergent (sat. soln.)	–	480	480	–	480	–	–
Bromine	G	–	–	–	–	G	G
Calcium hypochlorite	P	–	–	G	–	G	G
Carbon disulphide	P	NR	NR	G	24	P	F
Carbon tetrachloride[b]	P	NR	5	G	385	F	F
Chlorine	G	–	–	–	–	G	G
Chloroform[b]	P	NR	NR	G	9	F	P
Chromic acid	–	2	–	F	–	F	E
Cyclohexane	F	–	–	–	–	E	P
Cyclohexanol	–	1	163	–	480	–	–
Cyclohexanone	–	15	15	–	71	–	–
Dichloromethane[b]	–	NR	NR	G	5	F	F
Diethanolamine	F	–	–	–	–	E	E
Diethyl ether	–	1	3	E	62	G	P
Dimethyl formamide	–	15	29	–	28	–	–
Dimethyl sulphoxide[c]	–	–	–	–	–	–	–
Dioxane	–	1	8	–	NR	–	–
Ethanol	–	1	15	–	260	–	–
Ethanolamine	F	257	465	–	480	E	E
Ethylene glycol	G	480	480	E	480	G	E
Fluorine	G	–	–	–	–	G	G
Formaldehyde	G	30	30	E	480	E	E
Formic acid, 90%	G	21	69	E	78	E	E
Glycerol	G	–	–	E	–	G	E
Hexane	P	NR	NR	–	480	E	P
Hydrobromic acid (40%)	G	–	–	–	–	E	E
Hydrochloric acid (36%)	–	1	447	–	480	–	–
Hydrochloric acid (conc.)	G	–	–	G	–	G	E
Hydrofluoric acid (48%)	G	69	159	G	131	G	E
Hydrogen peroxide (100 vol.)	G	480	480	G	480	G	G
Iodine	G	–	–	–	–	G	G
Methanol	–	2	9	–	93	–	–
Methyl ethyl ketone	F	1	3	G	10	G	P
Methyl methacrylate	–	1	2	–	31	–	–
Morpholine	F	15	29	–	NR	E	E
Naphthalene[b]	G	–	–	E	–	G	G
Nitric acid (50%)	–	1	480	–	262	–	–
Nitric acid (conc.)	P	–	–	P	–	P	G
Nitrobenzene	–	15	15	–	NR	–	–
Perchloric acid	F	–	–	F	–	G	E
Petroleum ether (60–80)	–	NR	–	–	480	–	–

Table A.7 Continued

Chemical	Natural rubber			Nitrile	Neoprene	Vinyl
	light-weight	heavy-weight		heavy-weight		
Phenol (sat. soln)	G	4	39	–	NR	E	E
Phosphoric acid (85%)	G	480	480	–	480	E	E
Potassium hydroxide (50%)	G	429	480	G	480	G	E
Propan-2-ol	–	1	15	–	480	–	–
Sodium hydroxide (50%)	G	480	480	G	480	G	E
Sodium hypochlorite (sat. soln)	G	480	480	F	480	P	G
Sulphuric acid (50%)	–	480	480	–	180	–	–
Sulphuric acid (98%)	G	NR	–	F	180	G	G
Toluene[b]	–	NR	NR	G	44	F	F
1,1,1-Trichloroethane	–	NR	NR	–	133	–	–

[a]From ref. 2 and the Aldrich Catalogue. General chemical resistance for the material is given by a letter (E, excellent; G, good; F, fair; P, poor; NR, not recommended), while numerical entries are the times (minutes) for a chemical to pass through a specific type of glove with a particular thickness.
[b]Aromatic and halogenated hydrocarbons will attack all types of natural and synthetic glove materials. Should swelling occur, the user should change to fresh gloves and allow the swollen gloves to dry and return to normal.
[c]No data are given for the resistance of these materials to dimethyl sulphoxide. The manufacturer of the substance recommends the use of butyl rubber gloves.

Table A.8 Precipitation of toxic cations[a]

Group no.	Element[b] [precipitant][c]			
2	Be [OH⁻]	Ba [SO₄²⁻, CO₃²⁻]		
Lanthanides	Ln [OH⁻]			
4	Hf [OH⁻]			
5	V [OH⁻, S²⁻]			
6	Cr(III) [OH⁻]	Mo(VI) [d]	W(VI)[d]	
7	Mn(II) [OH⁻, S²⁻]	Re(VII) [S²⁻]		
8	Ru(III) [OH⁻, S²⁻]	Os(IV) [OH⁻, S²⁻]		
9	Co(II) [OH⁻, S²⁻]	Rh(III) [OH⁻, S²⁻]	Ir [OH⁻, S²⁻]	
10	Ni [OH⁻, S²⁻]	Pd [OH⁻, S²⁻]	Pt(II) [OH⁻, S²⁻]	
11	Ag [Cl⁻, OH⁻, S²⁻]			
12	Cd [OH⁻, S²⁻]	Hg [OH⁻, S²⁻]		
13	Ga [OH⁻]	In [OH⁻, S²⁻]	Tl [OH⁻, S²⁻]	
14	Sn [OH⁻, S²⁻]	Ge [OH⁻, S²⁻]	Pb [OH⁻, S²⁻]	
15	As [S²⁻]	Sb [OH⁻, S²⁻]	Bi [OH⁻, S²⁻]	
16	Se [S²⁻]	Te [S²⁻]		

[a]Metal cations precipitated as hydroxides or oxides at high pH. To prevent the precipitate from redissolving in excess base, it is often necessary to control pH, and ref. 2 gives recommended pH ranges for a range of cations. Sulphides are precipitated by addition of sodium sulphide to a neutral solution of the cation. Control of pH is important as some sulphides will redissolve in the presence of excess sulphide ion. After precipitation, excess sulphide is destroyed by addition of hypochlorite.
[b]Where indicated, the precipitant is for the particular oxidation state.
[c]Precipitants are listed in order of preference.
[d]These ions are best precipitated as CaMoO₄.

A.5 Chemical resistance of glove materials

Use Table A.7 to select the appropriate gloves when handling particular chemicals.

A.6 Disposal of residues

Air-sensitive or otherwise reactive residues should be decomposed safely before disposal. Alkali metals may be destroyed by isopropyl or *t*-butyl alcohol. Toxic cations may be removed from liquid waste by precipitation, as indicated in Table A.8.

General references

1. Urben, P.G., Pitt, M.J. and Battle, L.A. (eds) (1995) *Bretherick's Handbook of Reactive Chemical Hazards*, 5th edn, Butterworth-Heinemann, Oxford.
2. Lide, D.R. and Frederikse, H.P.R. (1996) *CRC Handbook of Chemistry and Physics*, 77th edn, CRC Press, Boca Raton, FL, USA.
3. Luxon, S.G. (ed.) (1992) *Hazards in the Chemical Laboratory*, 5th edn, The Royal Society of Chemistry, Cambridge.
4. Richardson, M.L. (ed.) (1992) *The Dictionary of Substances and their Effects*, Royal Society of Chemistry, Cambridge.

Appendix B   Deoxygenation columns

Supported copper metal and supported Mn(II) and Cr(II) oxides are the most common oxygen scavengers. A literature procedure for making highly active supported copper metal is available [1], but most researchers use the commercially available BTS Catalyst or, alternatively, prepare their own supported Mn(II) or Cr(II) columns as described below. The manganese system is capable of reducing oxygen levels to as low as 10^{-11} ppm and, unlike the chromium material, is not an active alkene polymerization catalyst and can therefore also be used to purify alkenes [2].

B.1   Supported MnO

The procedure below uses fine-mesh silica for ultrapurification of gases, and can be scaled up by using larger-mesh silica to give material suitable for packing larger columns [3].

Silica gel (60–70 mesh, 105 g) is washed with 1.5% nitric acid, then with distilled water and dried overnight at 100°C. A solution of $Mn(NO_3)_2$ (32 g) in distilled water (85 cm^3) is added in 3-cm^3 portions with stirring to the dried silica gel over about 1 h. The silica gel should appear dry when all of the solution has been added. Pack the silica gel into a glass tube (Figure 3.5) and heat the material overnight at 100°C to give a black solid, and then further oxidize this at 300°C in oxygen.

Note: Handle the silica gel in a fume cupboard to avoid breathing the dust. The nitrogen oxides produced in both of the heating steps should be absorbed in a base and vented to a fume cupboard.

Reduce the manganese oxide in a hydrogen stream at 370°C for 1 h to give a pale green, pyrophoric solid, which **must not** be exposed to a sudden inrush of air. As the MnO removes oxygen from the gas stream, a brown band of oxidized material extends down the column. When necessary, the column is regenerated by repeating the reduction step described above. Water produced during this process should be removed before incorporating the column into the gas supply again.

B.2 Supported CrO [4]

Dissolve CrO_3 (20 g) in distilled water (600 cm^3) and add silica gel (35–70 mesh, 400 g) with stirring. The orange mixture, which should appear almost dry, is further dried at 60°C *in vacuo* for about 8 h, or until the solid is a darker, sandy orange–brown.

Note: Cr(VI) compounds are **carcinogenic**. Wear suitable gloves and carry out operations in an efficient fume cupboard. Avoid breathing the dust from the silica gel.

Pack the material into a glass tube (Figure 3.5) and heat the column to 500°C in an oxygen stream for 30 min (the temperature should not be allowed to rise above 500°C, as the glass is close to its softening point). Turn off the furnace and pass nitrogen through the column for 5 min, then reduce the chromium in a stream of CO at 350°C for 15 min. Flush the column with nitrogen and allow it to cool. As with supported MnO, the blue CrO on silica should not be exposed to large volumes of air. The material turns from blue to brown upon oxidation, and is regenerated by heating in a hydrogen stream at about 400°C.

References

1. Meyer, F.R. and Ronge, G. (1939) *Angew. Chem.*, **52**, 637.
2. Moeseler, R., Horvath, B., Lindenau, D., *et al.* (1976) *Z. Naturforsch.*, **31B**, 892.
3. Shriver, D.F. and Drezdzon, M.A. (1986) *The Manipulation of Air-Sensitive Compounds*, 2nd edn, Wiley, New York, p.79.
4. Krauss, H.L. and Stach, H. (1969) *Z. Anorg. Allg. Chem.*, **366**, 34.

Appendix C Solvents

Physical properties of a range of solvents are given in Table C.1. Common names are listed in alphabetical order, and systematic names or synonyms are included underneath the primary name. Flash Point T_{FP} is defined as the lowest temperature at which a liquid emits vapour in sufficient concentration to form an ignitable mixture with air near the surface of the liquid, and the Autoignition Temperature T_{AI} is the minimum temperature required to initiate self-sustained combustion, regardless of heat source. Threshold Limit Values (TLV) are airborne concentrations (in ppm) which are believed that workers may be exposed to on a daily basis without adverse effects, and are given as time-weighted averages over an 8 h working day, unless followed by a C, in which case the value is a ceiling which should not be exceeded, even for short periods.

Table C.1 Solvent properties[a]

Solvent	m.p. (°C)	b.p. (°C)	ρ (g cm⁻³)	ε	T_c (°C)	P_c (MPa)	V_c (cm³ mol⁻¹)	T_{FP} (°C)	T_{AI} (°C)	TLV
Acetic acid (ethanoic acid)	17	118	1.04	6.15	320	5.786	171	43	427	10
Acetone (2-propanone)	-95	56	0.79	20.7	235	4.700	209	-19	538	750
Acetonitrile (ethanenitrile)	-44	82	0.78	36.2	272	4.85	173	6	524	40
Ammonia	-78	-33	0.68	25	132	11.35	72	–	–	25
Benzene	6	80	0.87	2.27	289	4.898	259	-11	560	0.1
Benzonitrile	-13	191	1.01	25.2	426	4.21	–	71	–	–
Butyl alcohol (1-butanol)	-90	118	0.81	17.1	290	4.423	275	29	345	50 C
t-Butyl alcohol (2-methyl-2-propanol)	25	82	0.78	10.9	233	3.973	275	11	478	100
Carbon disulphide	-112	46	1.26	2.64	279	7.90	173	-30	90	10
Carbon tetrachloride (tetrachloromethane)	-23	77	1.58	2.23	283	4.516	276	–	–	5
Chlorobenzene	-45	132	1.10	5.62	359	4.52	308	29	593	10
Chloroform (trichloromethane)	-64	61	1.48	4.81	263	5.47	239	–	–	10
Cyclohexane	7	81	0.77	2.02	280	4.07	308	-18	260	300
o-Dichlorobenzene	-17	180	1.30	9.93	–	–	–	65	648	25 C
1,2-Dichloroethane	-36	84	1.25	10.4	288	5.4	225	13	413	10
Dichloromethane	-95	40	1.32	8.9	237	6.10	–	–	600	50
Diethyl ether	-116	34	0.78	4.34	194	3.638	280	-45	160	400
Dimethoxyethane	-58	85	0.85	–	263	3.87	271	0	–	–
N,N-Dimethylformamide	-60	153	0.94	36.7	376	–	–	57	445	10
Dimethylsulphoxide	19	189	1.10	49.0	–	–	–	88	215	–
1,4-Dioxane	11	101	1.03	2.21	314	5.21	238	12	180	25
Ethanol	-114	78	0.79	24.3	241	6.132	167	12	363	1000
Ethyl acetate (ethyl ethanoate)	-84	77	0.89	6.02	250	3.882	286	-4	426	400
Ethylene glycol (1,2-ethanediol)	-13	197	1.11	37.7	449	–	–	111	400	50 C

Table C.1 Continued

Solvent	m.p. (°C)	b.p. (°C)	ρ (g cm^{-3})	ε	T_c (°C)	P_c (MPa)	V_c (cm^3 mol^{-1})	T_{FP} (°C)	T_{AI} (°C)	TLV
2-Methoxyethanol	−85	124	0.96	16.0	–	–	–	39	285	5
Heptane	−91	98	0.68	1.92	–	–	–	−4	204	400
Hexane	−95	69	0.66	1.89	235	3.010	370	−22	223	50
Hexamethylphosphoramide	7	235	1.03	30	–	–	–	105	–	–
Methanol	−98	65	0.79	32.6	239	8.092	118	11	385	200
Methylcyclohexane	−127	101	0.77	2.02	299	3.471	368	−4	285	400
Nitrobenzene	6	211	1.20	34.78	–	–	–	88	482	1
Nitromethane	−29	101	1.13	35.9	315	5.87	173	35	418	20
Octane	−57	126	0.70	1.95	296	2.493	492	13	206	300
Pentane	−130	36	0.62	1.84	197	3.364	311	−40	260	600
Propanenitrile (ethyl cyanide)	−93	97	0.78	27.2	288	4.26	229	2	512	–
1-Propanol	−126	97	0.80	20.1	291	5.168	219	15	370	200
2-Propanol (isopropyl alcohol)	−90	82	0.78	18.3	235	4.762	220	12	399	400
Pyridine	−42	115	0.98	12.3	347	5.67	243	20	482	5
Sulpholane	28	285	1.27	–	–	–	–	177	–	–
Tetrahydrofuran	−108	65	0.88	18.5	267	5.19	224	−14	321	200
Toluene	−95	111	0.86	2.38	319	4.104	316	4	530	50
1,3,5-Trimethylbenzene (mesitylene)	−45	165	0.86	2.28	364	3.127	–	50	550	25
Xenon	−112	−108	–	–	17	5.84	118	–	–	–
o-Xylene (1,2-dimethylbenzene)	−25	144	0.88	2.57	357	3.730	369	17	464	100
m-Xylene (1,3-dimethylbenzene)	−48	139	0.86	2.34	344	3.535	376	25	528	100
p-Xylene (1,4-dimethylbenzene)	13	138	0.86	2.27	343	3.511	379	25	529	100
Water	0	100	1.00	78.4	374	22.06	56	–	–	–

Appendix D   NMR Solvents

Rather than add a reference compound such as $SiMe_4$ to NMR samples, it is common practice to use solvent resonances as reference peaks, the protio residues in deuterated solvents providing suitable peaks for ^{1}H-NMR spectra. Table D.1 contains ^{1}H and ^{13}C chemical shifts for a range of NMR solvents, but it should be noted that these values can vary significantly with the nature and concentration of solute (up to several ppm for δ_C), so you may see different values quoted in other texts. If you need precise chemical shifts, it is therefore best to add a reference compound to your sample.

^{1}H-NMR spectra of three impurities which often find their way into NMR samples are shown in Figures D.1, D.2 and D.3. Silicone and Apiezon greases are leached from ground-glass joints and taps, while phthalate plasticizers are readily dissolved by organic solvents in contact with plastic tubing (e.g. Tygon or Portex). These spectra also show a peak due to water in the deuteriochloroform at about 1.6 ppm.

Table D.1   Properties of NMR solvents

Solvent	m.p. (°C)	b.p. (°C)	δ_H (ppm)	δ_C (ppm)	$^2J(^{13}C-^2H)$ (Hz)
Acetic acid-d_4	17	118	2.1	21.1	20
			(11.5)	(177.3)	
Acetone-d_6	−94	57	2.2	30.2	20
				(205.1)	0.9
Acetonitrile-d_3	−45	82	2.0	0.3	21
				118.2	
Benzene-d_6	5	80	7.3	128.7	24
Carbon tetrachloride	−23	77	−	96.7	
Chloroform-d	−64	62	7.26	77.7	32
Cyclohexane-d_{12}	6	81	1.4	27.8	19
Deuterium oxide	3.8	101.4	(4.7)	−	
1,2-Dichloroethane-d_4	−40	84	3.7	51.7	23.5
Dichloromethane-d_2	−95	40	5.3	54.2	27
Diethyl ether-d_{10}	−116	35	1.1	14.5	19
			3.3	65.3	21
Dimethylformamide-d_7	−61	153	2.7	30.1	21
			2.9	35.2	21
			8.0	162.7	21
Dimethylsulphoxide-d_6	18	189	2.6	43.5	21
Ethanol-d_6	−130	79	1.2	17.9	19
			3.7	57.3	22
Methanol-d_3	−98	65	3.5	49.3	21.5
Nitrobenzene	6	211	−	o 123.4	−
				m 129.4	−
				p 134.6	−
				ipso 148.2	−
Nitrobenzene-d_5			o 7.5	o 123.5	26
			m 7.67	m 129.5	25
			p 8.1	p 124.8	24.5
Nitromethane-d_3	−29	101	4.3	57.3	22
Pyridine-d_5	−42	116	α 8.7	α 149.8	27.5
			β 7.2	β 123.5	25
			γ 7.55	γ 135.5	24.5
Tetrahydrofuran-d_8	−109	66	1.8	26.7	20.5
			3.7	68.6	22
Toluene-d_8	−95	111	2.3	20.4	19
			o 7.1	o 128.9	23
			m 7.1	m 128.0	24
			p 6.98	p 125.2	24
				ipso 137.5	

D.1 ¹H-NMR spectra of common impurities

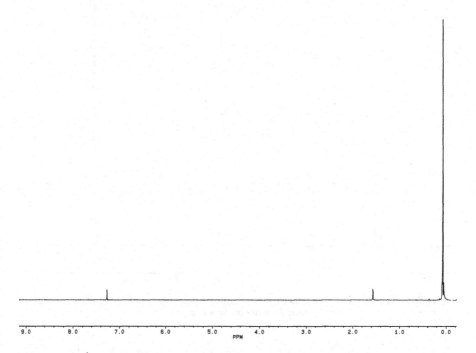

Figure D.1 ¹H-NMR spectrum (500 MHz, CDCl₃) of silicone grease. Peak at about δ1.6 is water impurity in the solvent.

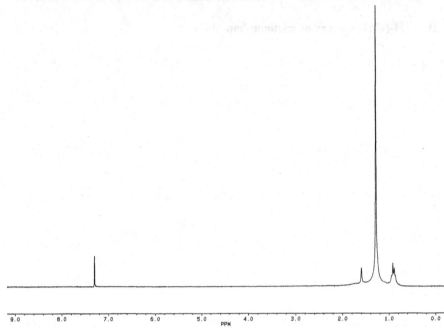

Figure D.2 ^{1}H-NMR spectrum (500 MHz, CDCl$_3$) of Apiezon grease. Peak at about $\delta1.6$ is water impurity in the solvent.

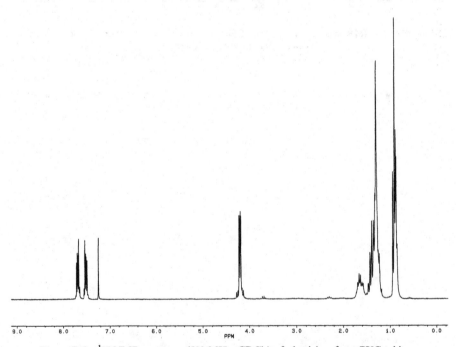

Figure D.3 ^{1}H-NMR spectrum (500 MHz, CDCl$_3$) of plasticizer from PVC tubing.

Appendix E Gases

Table E.1 Properties of gases

Gas	b.p. (°C)	m.p. (°C)	Density^a (g l^{-1})	Hazards	TLV (ppm)
Acetylene	−84		1.109	As, Fl, ▲	
Ammonia	−33	−78	0.71	To, Co, Fl	25
Argon	−189	−186	1.66	As	
Carbon dioxide		−78	1.83	Co	5000
Carbon monoxide	−191	−207	1.16	To, Fl	50
Chlorine	−34	−101	2.97	To, Co	1
Ethene	−104	−170	1.17	Fl	
Fluorine	−188	−220	1.57	Fl, Co	1
Helium	−269	−272	0.17	As	
Hydrogen	−253		0.08	Fl	
Hydrogen bromide	−67	−87	3.34	To, Co	3
Hydrogen chloride	−85	−114	1.52	To, Co	5
Hydrogen fluoride	19.5	−83	0.94	To, Co	3
Hydrogen sulphide	−60	−85	1.43	To, Co	10
Nitric oxide (NO)	−152	−164	1.24	To	25
Nitrogen	−196		1.25	As	
Nitrogen dioxide (NO$_2$)	21	−9.3	3.3	To, Co	3
Oxygen	−183	−218	1.33	Ox	
Sulphur dioxide	−10	−75	2.70	To, Co	2

^{a}Density of gas
As, asphyxiant; Co, corrosive; Fl, flammable; Ox, oxidant; To, toxic; ▲, potentially explosive when pressurized.

Index

Page numbers for figures are in **bold** and page numbers for tables are in *italics*.

Printed in the United States
by Baker & Taylor Publisher Services